Nanotechnology

Nanotechnology

Science, Innovation, and Opportunity

Lynn E. Foster

Upper Saddle River, NJ • Boston • Indianapolis • San Francisco
New York • Toronto • Montreal • London • Munich • Paris • Madrid
Capetown • Sydney • Tokyo • Singapore • Mexico City

Library of Congress Cataloging-in-Publication Data

Foster, Lynn E.
Nanotechnology : science, innovation and opportunity / Lynn E. Foster.
p. cm.
Includes bibliographical references and index.
ISBN 0-13-192756-6 (hard cover : alk. paper)
1. Nanotechnology. 2. Nanotechnology--Social aspects. I. Title.
T174.7.F68 2005
620'.5—dc22

2005027597

ISBN: 0-13-192756-6
Text printed in the United States on recycled paper at Courier in Westford, Massachusetts.
First printing, December 2005

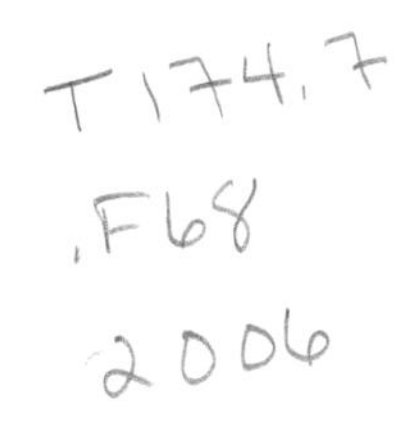

Dedicated to Richard Feynman:
For having a vision of things smaller than any of us
and a spirit bigger than all of us.

And to my wife Miriam:
For enduring the vision of one man.

Contents

Foreword

When Congress passed and the President signed the 21st Century Nanotechnology Research and Development Act, or Public Law 108-153, in December 2003, our goal was to help spur and coordinate research and technology development in this exciting and promising area, one that has the potential to transform every aspect of our lives. By manipulating matter at a molecular scale, nanotechnology may allow us to develop new materials and devices that have unique properties beyond the realm of current conventional technology. Its potential economic effects are also profound, as some have estimated that its impact on both existing and new industries may easily exceed tens of billions of dollars by the end of this decade, and a trillion dollars not too many years beyond. Applications ranging from novel and highly targeted therapeutic treatments for diseases including cancer to such timely concerns as increased energy efficiency or tools for providing a cleaner environment certainly emphasize the potentially transformational nature of nanotechnology.

While many potential applications are some years ahead of us, the increasing frequency with which articles on nanotechnology are appearing in trade publications, investment publications, and in the mainstream media tell us that the science is already moving beyond the laboratory. Yet as we move ahead with the development of this frontier technology, it is important that we also pay attention to the issues beyond fundamental science and discovery that are equally important—issues such as technology innovation and technology transfer. And it is not just economic issues that need to be considered. As with any new technology, we need to pay careful attention to minimizing any potential impact on the environment as we move ahead. As the chapters in this book indicate, while much work lies ahead, we are making exciting progress in each of these areas.

Senator Joe Lieberman
Senator George Allen
Washington, D.C., November 2005

Preface

During the past century, human life spans have almost doubled, and travel and communication happen with an ease and speed that would have been considered science fiction only a few generations ago. Remarkably, the pace of innovation is actually increasing over that of the past.

Science has now advanced to the point that those on the cutting edge of research work with individual atoms and molecules. This is the defining characteristic of the new metafield of nanotechnology, which encompasses a broad range of both academic research and industrial development. At this small scale, the familiar classical physics guideposts of magnetism and electricity are no longer dominant; the interactions of individual atoms and molecules take over. At this level—roughly 100 nanometers (a nanometer being a billionth of a meter, and a human hair being 50,000 nanometers wide) and smaller—the applicable laws of physics shift as Newtonian yields to quantum.

Nanotechnology holds the promise of advances that exceed those achieved in recent decades in computers and biotechnology. Its applications will have dramatic infrastructural impacts, such as building tremendously faster computers, constructing lighter aircraft, finding cancerous tumors still invisible to the human eye, or generating vast amounts of energy from highly efficient solar cells. Nanotechnology will manifest in innovations both large and small in diverse industries, but the real benefit will accumulate in small cascades over decades rather than in a sudden, engulfing wave of change. It is not the "Next Big Thing" but rather will be any number of "next large things". Nanotechnology may not yield a result as dramatic as Edison's lightbulb but rather numerous gains as pervasive as the integrated-circuit-controlled lightbulbs in the traffic lights that are ubiquitous in modern life.

Although the lightbulb breakthroughs will be few, there will be numerous benefits taken for granted, such as the advantages that the automated intelligence of traffic grids provide to major cities. This should not be a surprise, because nanotechnology is not an invention but rather a range of fields of study and applications, defined by size, that use tools, ideas, and intuitions available to innumerable scientific disciplines. Thus nanotechnology offers tremendous potential for several key reasons. Materials and processes at that size have unique properties not seen at larger scale, offer proportionately greater reactive surface area than their larger counterparts, and can be used in or with living organisms for medical applications. As a result, familiar materials can have completely different properties at the nanoscale.

For example, carbon atoms form both coal and diamonds, but with different molecular arrangements. Scientists now know that carbon molecules at the nanoscale can form cylindrical tubes, called carbon nanotubes, that are much stronger than steel and conduct electricity, neither of which is possible with the carbon found in coal or diamonds. Carbon nanotubes may one day provide key breakthroughs in medicine and electronics. Likewise, nanotechnology can provide breakthroughs in industrial uses. The electrical current produced in solar cells or batteries reflects the flow of electrons from one surface to another. Nanotechnology has already enabled the demonstration of a vastly increased surface area of electrodes that allows electrons to flow much more freely, along with corresponding improvements in battery performance. Safer, cheaper, and cleaner electricity and electrical storage would obviously have a dramatic impact on our society.

Another reason nanotechnology holds so much promise is that it enables solutions at the same size scale as biological organisms, such as the individual cells in our bodies. Engineered materials are possible, such as ultrasmall particles made in the exact size to perform like a "smart bomb" in delivering drugs in the blood stream. Other applications might detect cancer when it is only a few cells in size. Future convergence of nanotechnology and biotechnology may combine biological and man-made devices in a variety of applications, such as batteries for implanted heart pacemakers that draw electrical current from the wearer's glucose rather than from surgically implanted batteries.

Yet another important facet of nanotechnology—one that underpins both its promise and the challenges—is that it embraces and attracts so many different disciplines that researchers and business leaders are working in, among them, chemistry, biology, materials science, physics, and computer science. Although each field has tremendously talented people, each also has its own somewhat unique training and terminology. Almost like the parable of the blind men and the elephant, each group approaches the molecular level with unique skills, training, and language. Communication and research between academic disciplines and between researchers and their business counterparts is critical to the advancement of nanotechnology.

With the diversity of professional cultures in mind, a central goal of this book is to promote communication and cooperation between researchers and industry by including similarly diverse articles written by experts but accessible to everyone.

The depth of scientific talent and the substantial resources being devoted to nanotechnology are a tremendous cause for optimism for both near-term and long-term gains. Ultimately nanotechnology will yield greater impact than information technology or biotechnology has. However, the tempo of technology is not set by the velocity of novel discoveries, but rather by the pace of what the market will embrace and pay for. The medium term in nanotechnology will be difficult and delayed by issues far beyond scientific research or product prototyping—namely, by the long, difficult process of new products gaining traction in the marketplace. To reach the stage of a viable product, the innovations will have to overcome issues such as how they are integrated, how much power they consume, and how they are controlled. Only then will the marketplace vote with dollars on the technology. For these reasons, another goal of this book is to highlight these issues so that a broader audience can address them with its respective understanding and resources.

This book is organized into four matrixed sections. Section One is focused on the history and development drivers of innovation. The first chapter highlights a historical example from the early days of the biotechnology industry as a cautionary lesson about a new industry developing with new tools and tremendous promise. The promise of nanotechnology to solve the world's energy problem is outlined in Chapter 2, along with the impact the solution would have on solving other problems as well. Chapter 3 is a discussion of the role played by expectations in the development of an industry.

Section Two focuses on the talents, roles, and motivations of the main players and individuals, along with the organizational factors that drive technologies forward or limit their impact. Chapter 4 presents the vision of a venture capitalist who takes a long-term view of nanotechnology as the nexus of disruptive innovation, and Chapter 5 outlines current investment decisions in nanotechnology. Chapter 6 outlines the U.S. government's role in funding research and establishing policies for the safe and effective use of nanotechnology. Then Chapter 7 discusses specific areas of academic research, and Chapter 8 explains how technologies developed there are brought to commercial use. The role of U.S. patent law in commerce follows in Chapter 9, with a discussion of its impact on the advance of nanotechnology. Chapter 10 explains why entrepreneurs are the key drivers of change in a new industry and help it advance by taking tremendous personal risks. Chapter 11 discusses the challenges within a large corporation that is developing technology products.

Finally, Chapter 12 presents an overview of technologies developed in federal laboratories and describes how they are commercialized.

Section Three considers specific areas of innovation: nanoscale materials (Chapter 13) as well as other areas where nanotechnology is making a dramatic impact: nano-enabled sensors (Chapter 14), the microelectronics industry (Chapter 15), and drug delivery (Chapter 16). This part concludes with a chapter (Chapter 17) specifically on the intersection of nanotechnology and biotechnology, a combination that holds enormous potential to impact medicine and health.

Section Four suggests that the convergence of science at the nanoscale foreshadows a transformation and revolutionary change in society (Chapter 18) and highlights ethical considerations in the advance of nanotechnology (Chapter 19).

The Epilogue features a prescient speech given in 1983 by the late Richard Feynman, the legendary physicist who first envisioned nanotechnology.

Working at the level of individual atoms and molecules allows researchers to develop innovations that will dramatically improve our lives. The new realm of nanotechnology holds the promise of improving our health, our industry, and our society in ways that exceed even those of computers or biotechnology.

Acknowledgments

These acknowledgments must begin with, and I cannot thank enough, distinguished scientist Dr. Hans Coufal of IBM and the Nanoelectronics Research Initiative for the inspiration and encouragement to undertake this project.

Innumerable thanks go to the many authors who contributed to this volume. The responsiveness and generosity of such a distinguished group of researchers and business professionals transformed the arduous task of bringing this book together into a particularly gratifying one. Special thanks are also due to Bernard Goodwin of Prentice Hall for his continuing guidance.

Behind my efforts is one person whose day-to-day patience and encouragement were essential: my wife, Miriam, who has shared the highs and lows of a consuming project. These few words are only a small fraction of my enormous sense of love, appreciation, and thanks.

Four people worked "above and beyond" on this book, and without their special behind-the-scenes efforts it would not have been completed: Michael Krieger of Williken Wilson Loh & Stris LLP, Geoff Holdridge, National Nanotechnology Coordination Office and WTEC, Inc., Matt Laudon of the Nano Science & Technology Institute, and Brent Segal of Nantero.

Thanks are due also to several people who contributed greatly to this book but were not listed as authors because of editorial constraints, in particular the following: Art Riggs, City of Hope's Beckman Research Institute; Howard Schmidt, Rice University; Amit Kenjale, Caltech; Robert Giasoli, XCOM Wireless and the Micro and Nanotechnology Commercialization Education Foundation; Jim Gimzewski, University of California, Los Angeles, and the California NanoSystems Institute; Michael Holton, Smart Technology Ventures; Jim Spohrer, IBM; Daniel Choi, NASA Jet Propulsion Lab; Marc Madou, University of California, Irvine; Peggy Arps, University of California, Irvine; Susan Zhou, University of California, Irvine; Tony Cheetham, University of California, Santa Barbara.

Numerous colleagues at Greenberg Traurig, LLP, were a source of information, ongoing support, and encouragement, as was the firm itself. My nanotechnology universe was broadened by the opportunity to do programs for the Caltech Enterprise Forum, and by the wisdom and camaraderie of many of its board members.

Thanks for the foundation of this book are due to Dr. Joe Lichtenhan of Hybrid Plastics, who started me on the nanotechnology road, and to Jerry Gallwas, Hans Coufal, Malcolm Green, and Sergeant Major (Retired) Peter Van Borkulo for being mentors to me for so many years. Special appreciation goes to Cheryl Albus of the National Science Foundation for her constructive skepticism that helped me separate substance from fad in nanotechnology.

Many others contributed greatly and are due wholehearted thanks for helping bring this book to completion: Kevin Ausman, Rice University; John Belk, The Boeing Company; Keith Boswell, Virginia Economic Development Partnership; Peter Burke, University of California, Irvine; Altaf Carim, U.S. Department of Energy; Scott Carter, Caltech; Gerard Diebner, Greenberg Traurig; Andre Dehon, Caltech; Alex Dickenson, Luxtera; Sverre Eng, NASA Jet Propulsion Lab (retired); David Forman, Small Times Media; Wayne Jones, Greenberg Traurig; Sam Gambhir, Stanford; Guanglu Ge, Caltech; Maurice Gell, University of Connecticut; Peter Gluck, Greenberg Traurig; Peter Grubstein, NGEN Partners; Joe Helble, Office of Senator Joseph Lieberman and Dartmouth College; Jamie Hjort, Office of Senator George Allen; Julie Holland, NASA Commercialization Center, California Polytechnic University; Robert Hwang, U.S. DOE Brookhaven National Laboratory; Srinivas Iyer, Los Alamos National Laboratory; Barbara Karn, U.S. Environmental Protection Agency; David Knight, Terbine, Inc.; David Lackner, NASA Ames; Tony Laviano, Loyola Marymount University; Tom Loo, Greenberg Traurig; Goran Matijasevic, University of California, Irvine; Paul McQuade, Greenberg Traurig; John Miller, Arrowhead Research; Brian Pierce, Rockwell Scientific; Andy Quintero, The Aerospace Corporation; Phil Reilly, Macusight, Inc.; Tony Ricco, Stanford; Bart Romanowicz, The Nano Science & Technology Institute; Ari Requicha, USC; Michael Roukes, Caltech; James Rudd, The National Science Foundation; Frank Sadler, Office of Senator George Allen; Carlos Sanchez, U.S. Army; John Sargent, U.S. Department of Commerce; Jeff Schloss, National Institutes of Health; Clayton Teague, National Nanotechnology Coordination Office; Jim Von Ehr, Zyvex; Kumar Wickramasinghe, IBM; Stan Williams, HP; Hiroshi Yokoyama, National Institute of Advanced Industrial Science and Technology, Japan; CJ Yu, Glyport, Inc.

—Lynn Foster
Los Angeles, July 2005

About the Author

Lynn E. Foster is the Emerging Technologies Director of Greenberg Traurig, LLP, one of the largest law firms in the United States. In this position he advises technology companies on technology transfer, patent licensing, strategic partnerships and raising capital.

Prior to joining Greenberg Traurig, Mr. Foster held technology industry positions in corporate, entrepreneurial, and government settings, among them managing software development in the aerospace industry, heading a startup, and managing a seed-stage commercialization grant program. He serves on Advisory Boards for the Nano Science and Technology Institute and the International Engineering Consortium, as well as the Executive Committee of the Caltech Enterprise Forum. He authored the first Nanotechnology Trade Study and has directed eight Nanotechnology conferences and trade missions. He also has 20 years of active and reserve service with the U.S. Army, including service in the first Gulf War and Bosnia. He holds an M.B.A. and a B.S. in production and operations management.

Contributors

Gerald Gallwas

Gerald Gallwas was a member of the original team in the mid-1960s that founded and managed the growth of what became the clinical diagnostic business of Beckman Instruments. The team pioneered a new technology based on kinetic rate measurements applied to first-order chemical reactions for clinically significant blood constituents.

As the business grew, he served in many roles, from new product development to directing clinical field trials in the United States, Europe, and Japan. This led to extensive involvement with professional and trade organizations as well as regulatory agencies. He retired after 30 years of service as Director of Program Management, overseeing multimillion-dollar new product development programs. Since that time, he has consulted widely in the United States and Europe on new product development, strategic and business planning, organizational development, and marketing. Gallwas holds a B.S. in chemistry from San Diego State University.

Richard Smalley

Nobel Laureate Richard E. Smalley received his B.S. in 1965 from the University of Michigan. After an intervening four-year period in industry as a research chemist with Shell, he earned his M.S. in 1971 from Princeton University and his Ph.D. in 1973. During a postdoctoral period with Lennard Wharton and Donald Levy at the University of Chicago, Smalley pioneered what has become one of the most powerful techniques in chemical physics: supersonic beam laser spectroscopy. After coming to Rice University in 1976 he was named to the Gene and Norman Hackerman Chair in Chemistry in 1982. He was a founder of the Rice Quantum Institute in 1979 and served as the Chairman from 1986 to 1996. In 1990 he became a professor in the

Department of Physics and was appointed University Professor in 2002. He was the founding director of the Center for Nanoscale Science and Technology at Rice from 1996 to 2002, and is now Director of the new Carbon Nanotechnology Laboratory at Rice.

In 1990 he was elected to the National Academy of Sciences, and in 1991 to the American Academy of Arts and Sciences. He is the recipient of the 1991 Irving Langmuir Prize in Chemical Physics, the 1992 International Prize for New Materials, the 1992 E.O. Lawrence Award of the U.S. Department of Energy, the 1992 Robert A. Welch Award in Chemistry, the 1993 William H. Nichols Medal of the American Chemical Society, the 1993 John Scott Award of the City of Philadelphia, the 1994 Europhysics Prize, the 1994 Harrison Howe Award, the 1995 Madison Marshall Award, the 1996 Franklin Medal, the 1996 Nobel Prize in Chemistry, the Distinguished Public Service Medal awarded by the U.S. Department of the Navy in 1997, the 2002 Glenn T. Seaborg Medal, and the 2003 Lifetime Achievement Award of *Small Times* magazine. He received three honorary degrees in 2004: an Honorary Doctorate from the University of Richmond; a Doctor Scientiarum Honoris Causa from Technion Israel Institute of Technology, and a Doctor of Science from Tuskegee University.

Smalley is widely known for the discovery and characterization of C60 (Buckminsterfullerene, aka the "buckyball"), a soccer ball–shaped molecule that, together with other fullerenes such as C70, now constitutes the third elemental form of carbon (after graphite and diamond). His current research is on buckytubes: elongated fullerenes that are essentially a new high-tech polymer, following on from nylon, polypropylene, and Kevlar. But unlike any of these previous wonder polymers, these new buckytubes conduct electricity. They are likely to find applications in nearly every technology where electrons flow. In February 2000 this research led to the start up of a new company, Carbon Nanotechnologies, Inc., which is now developing large-scale production and applications of these miraculous buckytubes.

Peter Coffee

Peter Coffee is Technology Editor of *eWEEK,* Ziff Davis Media's national news magazine of enterprise infrastructure. He has twenty years' experience in evaluating information technologies and practices as a developer, consultant, educator, and internationally published author and industry analyst.

Coffee writes product reviews, technical analyses, and his weekly "Epicenters" column on disruptive forces in IT tools and practices; he has appeared on CBS, NBC, CNN, Fox, and PBS newscasts addressing Internet security, the Microsoft antitrust case, wireless telecom policies, and other e-business issues. He chaired the four-day Web Security Summit conference in Boston during the summer of 2000 and has been a keynote speaker or moderator at technical conferences throughout the United States and in England. His most recent book is *Peter Coffee Teaches PCs,* published in 1998 by Que; he previously authored Que's ZD Press tutorial, "How to Program Java." He played a lead role in developing *eWEEK* Labs' 2001 series of special reports titled "Five Steps to Enterprise Security." His current *eWEEK* beats include development tools and business intelligence products.

Before joining *eWEEK* (then called *PC Week*) full time in 1989, Coffee held technical and management positions at Exxon and The Aerospace Corporation, dealing with chemical facility project control, Arctic project development, strategic defense analysis, end-user computing planning and support, and artificial intelligence applications research. He has been one of *eWEEK*'s lead analysts throughout the life cycles of technologies, including x86 and RISC microprocessors; Windows, OS/2, and Mac OS; object technologies, including Smalltalk, C++, and Java; and security technologies including strong encryption. He holds an engineering degree from MIT and an M.B.A. from Pepperdine University and has taught classes in the Department of Computer Science at UCLA and at Pepperdine's Graziadio School of Business and Management and the Chapman College School of Business. His weekly newsletter, *Peter Coffee's Enterprise IT Advantage,* and his other writings are available at www.eweek.com/petercoffee.

Steve Jurvetson

Steve Jurvetson is a Managing Director of Draper Fisher Jurvetson (DFJ.com). He was the founding VC investor in Hotmail (MSFT), Interwoven (IWOV), and Kana (KANA). He also led the firm's investments in Tradex and Cyras (acquired by Ariba and Ciena for $8 billion) and, most recently, in pioneering companies in nanotechnology and molecular electronics. Previously, Jurvetson was an R&D Engineer at Hewlett-Packard, where seven of his communications chip designs were fabricated. His prior technical experience also includes programming, materials science research (TEM atomic imaging of GaAs) and computer design at HP's PC Division, the Center for Materials Research, and Mostek. He has also worked in product marketing at Apple and NeXT Software.

At Stanford University, he finished his B.S.E.E. in two and a half years and graduated first in his class, as the Henry Ford Scholar. Jurvetson also holds an M.S. in electrical engineering from Stanford. He received his M.B.A.

from the Stanford Business School, where he was an Arjay Miller Scholar. Jurvetson also serves on the Merrill Lynch and STVP advisory boards and is Co-Chair of the NanoBusiness Alliance. He was honored as "The Valley's Sharpest VC" on the cover of *Business 2.0* and was chosen by the *San Francisco Chronicle* and *San Francisco Examiner* as one of "the ten people expected to have the greatest impact on the Bay Area in the early part of the 21st Century." He was profiled in the *New York Times Magazine* and featured on the cover of *Worth* and *Fortune* magazines. He was chosen by *Forbes* as one of "Tech's Best Venture Investors," by the *VC Journal* as one of the "Ten Most Influential VCs," and by *Fortune* as part of its "Brain Trust of Top Ten Minds."

Daniel V. Leff

Prior to joining Harris & Harris Group, Leff was a Senior Associate with Sevin Rosen Funds in the firm's Dallas, Texas, office, where he focused on early-stage investment opportunities in semiconductors, components, and various emerging technology areas. While at Sevin Rosen Funds, he played an integral role in the funding of Nanomix, InnovaLight, Sana Security, and D2Audio. Leff has also worked for Redpoint Ventures in the firm's Los Angeles office. In addition, he previously held engineering, marketing, and strategic investment positions with Intel Corporation.

Leff received his Ph.D. in physical chemistry from UCLA's Department of Chemistry and Biochemistry, where his thesis adviser was Professor James R. Heath (recipient of the 2000 Feynman Prize in Nanotechnology). Leff also received a B.S. in chemistry from the University of California, Berkeley, and an M.B.A. from The Anderson School at UCLA, where he was an Anderson Venture Fellow. Leff has published several articles in peer-reviewed scientific journals and has been awarded two patents in the field of nanotechnology. He is also a member of the business advisory boards of the NanoBusiness Alliance and the California NanoSystems Institute (CNSI).

R. Douglas Moffat

R. Douglas Moffat, CFA, is President of Moffat Capital, LLC, and Nanotech Financing Solutions, LLC. Moffat Capital, LLC, is a broker/dealer experienced in raising private capital, and Nanotech Financing Solutions, LLC, is a leading consultancy in this emerging field.

Moffat has 30 years' experience as a research analyst at leading Wall Street brokerage firms and has broad experience in industrial markets. He published research on the metals and steel industries, diversified industrial companies, automotive suppliers, defense, engineering, industrial service, industrial technology sectors, and nanotechnology. An electrical engineer and M.B.A. by education, Moffat brings a critical technical viewpoint to identifying business models with merit.

Moffat is a Chartered Financial Analyst (CFA) and was selected five times a *Wall Street Journal* All-Star Analyst. He published on more than 110 public companies, with competency in the nanoscience, industrial, metals, power technology, and real estate sectors. He is a leading industry consultant in nanotechnology. Moffat is also a director of Astec Industries, Inc., a public company producing asphalt, road building, and construction aggregate equipment.

Geoffrey M. Holdridge

Geoffrey M. Holdridge is currently Vice President for Government Services at WTEC, Inc., a nonprofit corporation that provides on-site contract staff services for various U.S. government agencies as well as an ongoing series of international technology assessment studies for the government, comparing the status and trends in research and development in specific science and technology research areas overseas relative to the United States (see http://www.wtec.org). Holdridge's primary assignment for WTEC currently is to serve as Policy Analyst and Contract Staff Manager at the National Nanotechnology Coordination Office, which provides technical and administrative support to the Nanoscale Science, Engineering, and Technology (NSET) Subcommittee of the President's National Science and Technology Council (NSTC). NSET is the interagency body charged with coordinating, planning, and implementing the U.S. National Nanotechnology Initiative (see http://www.nano.gov). Holdridge also plays a similar role in providing staff support for NSTC's Multi-Agency Tissue Engineering Science (MATES) Working Group, which coordinates tissue science and engineering (also known as regenerative medicine) research across several federal agencies (http://www.tissueengineering.gov). Previous assignments for WTEC include Vice President for Operations, where he coordinated worldwide activities for WTEC mostly related to its series of international technology assessment studies, and Director of the Japanese Technology Evaluation Center (JTEC), with responsibility for studies assessing Japanese research and development relative to the United States. In the course of these assignments, Holdridge has supervised editing staff at WTEC and NNCO and has performed final editing passes on more than 70 WTEC and NNCO reports, including several that were formerly cleared for publication by NSTC under the seal of the President of the United States.

Prior to coming to WTEC, Holdridge served in several capacities at the National Science Foundation (NSF), in the Division of Policy Research and Analysis, the Division of Electrical and Communications Systems, and the Division of Emerging Engineering Technologies, where he was involved in helping organize NSF's Emerging Engineering Technologies Initiative. In the Division of Policy Research and Analysis he conducted internal studies at the request of the Office of Management and Budget and the Office of Science and Technology Policy and supervised extramural NSF funding for policy research on environmental, energy, and resources issues, with particular focus on renewable energy and energy conservation research. In the Division of Electrical and Communications Systems he served as an NSF representative to an interagency committee that helped coordinate federal R&D policy related to the semiconductor agency.

Holdridge also has held positions at the National Research Council of the National Academies and at the University of Virginia. He holds a bachelor's degree from Yale University in history.

Julie Chen

Julie Chen is currently the Director of the Nanomanufacturing Center of Excellence at the University of Massachusetts Lowell, where she is a professor of mechanical engineering and is also co-Director of the Advanced Composite Materials and Textile Research Laboratory. Chen was the Program Director of the Materials Processing and Manufacturing and the Nanomanufacturing Programs in the Division of Design, Manufacture, and Industrial Innovation at the National Science Foundation from 2002 to 2004.

She received her Ph.D., M.S., and B.S. in mechanical engineering from MIT. Chen has been on the faculty at Boston University, a NASA-Langley Summer Faculty Fellow, a visiting researcher at the University of Orleans and Ecole Nationale Superieure d'Arts & Metiers (ENSAM-Paris), and an invited participant in the National Academy of Engineering, Frontiers of Engineering Program. In addition to co-organizing several national and international symposia and workshops on nanomanufacturing and composites manufacturing for NSF, ASME, ASC, and ESAFORM, Chen has served on editorial boards, advisory committees, and review panels for several journals and federal agencies. Her research interests are in the mechanical behavior and deformation of fiber structures, fiber assemblies, and composite materials, with an emphasis on nanomanufacturing and composites processing. Examples include analytical modeling and novel experimental approaches to electrospinning of nanofibers, and forming of textile reinforcements for structural (biomedical to automotive) applications.

Larry Gilbert

Lawrence Gilbert is the Senior Director of Technology Transfer of Caltech, the California Institute of Technology. The Office of Technology Transfer (OTT) provides a service to Caltech faculty members, other Caltech researchers, and Jet Propulsion Laboratory (JPL) technologists by protecting the intellectual property developed in their Caltech and JPL labs. The OTT fosters the commercial development of Caltech technologies in companies ranging from local start-ups to large, multinational firms. Gilbert has been responsible for the formation of more than 60 start-ups based upon or associated with university research. Several have gone public or have been acquired, and many have products in the marketplace. He acts primarily as a catalyst in putting the deal together, linking faculty, technology, and venture capital.

Gilbert was formerly the Director of Patent Licensing for Massachusetts Institute of Technology (MIT). Prior experience includes Patent Consultant to various universities, including Boston University, Brandeis, Tufts, and the University of Massachusetts Medical Center and as the Director of Patent and Technology Administration of Boston University. Gilbert is a member of the Licensing Executive Society (LES) and a former Chairman of its Committee on Technology Transfer, and is a Founder of the former Society of University Patent Administrators, now known as the Association of University Technology Managers (AUTM). He is a member of the Executive Committee of the MIT/Caltech Enterprise Forum and formerly a member of the Board of Directors of the Southern California Biomedical Council and a former member of the Advisory Committee of the Business Technology Center, a high-tech incubator sponsored by the Los Angeles County Community Development Commission.

Gilbert received his B.A. from Brandeis University; an M.I.M. degree from the American Graduate School of International Management (Thunderbird); and a J.D. from Suffolk University; he is registered to practice before the U.S. Patent and Trademark Office (USPTO). Gilbert has been a frequent lecturer on patent and licensing matters and has written several articles in the field.

Michael Krieger

Michael Krieger is a Los Angeles attorney with Willenken Wilson Loh & Stris. He has practiced high-technology business and intellectual property law for more than 20 years, focusing on strategic counseling, litigation strategy, and preventive methods to both secure and exploit clients' key IP assets. This includes both patent, trademark, and other litigation and associated transactions for development, acquisition, licensing, and services. His clients have ranged from start-ups to industry leaders, the United Nations, and international Web-based initiatives. He also has been an expert in technology litigation.

With degrees in mathematics (B.S., Caltech; Ph.D., UCLA) and law (UCLA), Krieger was on the MIT Mathematics and UCLA Computer Science faculties as well as a Fulbright Scholar prior to practicing law. Combining knowledge of both law and new technology, he was involved early on with emerging issues such as public key encryption (1978 tutorial for the Computer Law Association), domain name/trademark conflicts, open source software, and Internet governance (for example, personal adviser to Internet pioneer Jon Postel as he oversaw privatization of Internet administration).

Krieger serves on several boards supporting new venture development, including the Caltech/MIT Enterprise Forum. In 1997 he developed and continues to teach a graduate seminar on entrepreneurial business, IP, and litigation issues for the Computer Science Department in UCLA's engineering school. He has served on the Executive Committee of the Intellectual Property Section of the California Bar, was Editor-in-Chief of its journal *New Matter*, and currently chairs its Computer Law Committee.

Chinh H. Pham

Chinh H. Pham practices law at Greenberg Traurig, LLP (www.gtlaw.com), and heads the firm's nanotechnology practice. He is a registered patent attorney with particular experience in the strategic creation, implementation, and protection of intellectual property rights for high-technology and life science clients.

Chinh represents and counsels start-ups as well as established companies, including those in the areas of nanotechnologies, medical devices, electromechanical devices, telecommunications, data mining, electronic commerce, and life sciences. In connection with his practice, Chinh advises clients on the creation and development of patent portfolios through the preparation and filing of patent applications, the acquisition and exploitation of intellectual property rights through licensing and strategic collaboration agreements, and the preparation of invalidity, noninfringement, and freedom-to-operate opinions. Chinh also counsels clients on IP due diligence through the evaluation of client and competitor portfolios.

For his start-up clients, Chinh assists with strategies for leveraging their IP portfolio for high-value commercial opportunities, introducing them to funding sources, either through the venture community or the government, as well as identifying and establishing for these clients strategic alliances with industry leaders. Chinh is the founder of the NanoTechnology and Business Forum, a monthly forum dedicated to the business of nanotechnology. He is also the Secretary of ASTM International Nanotechnology Committee. Chinh is a frequent speaker and writer on the intersection between nanotechnology, intellectual property, and business.

Prior to practicing law, Chinh worked in the biotechnology industry, focusing on the genetic transformation of plants. Chinh received his B.A. in genetics from the University of California, Berkeley, and his J.D. from the University of San Francisco.

Charles Berman

Charles Berman is a principal shareholder with the law firm Greenberg Traurig, LLP, in Santa Monica, California, and he is Co-Chair of the national patent prosecution practice of the firm. He has been active in all areas of intellectual property law for more than 35 years. He has been practicing in Los Angeles since 1978, and practiced in South Africa prior to that time. He has also worked with European patent attorneys.

Berman has extensive IP experience in the United States and foreign countries and has been involved in the U.S. and international licensing of technology. Berman has represented large corporations in actions before the U.S. Patent Office, the European Patent Office, and the German, Japanese, and other major patent offices. He has engaged in major international patent and trademark procurement programs for leading corporations.

Berman currently serves as Chair of Fellows of the American Intellectual Property Law Association (AIPLA), where he was a founding fellow, and has also served on the board of directors. He has held several prominent positions over the past several years in the AIPLA, including being a representative for international meetings. He has been President of the Los Angeles Intellectual Property Law Association.

He is a frequently invited speaker, has written numerous articles on the topic of intellectual property, and has been on the editorial boards of different intellectual property publications. Berman was educated at the University of Witwatersrand in South Africa.

Jeff Lawrence

Jeff Lawrence is President & CEO of Clivia Systems (www.cliviasystems.com), former CTO of Intel's Network Communications Group, and co-founder and former President and CEO of Trillium Digital Systems. Trillium was a leading provider of communications software solutions to communications equipment manufacturers building the converged wireless, Internet, broadband, and telephone infrastructure. Trillium licensed its source code software solutions to more than 300 companies throughout the world for use in a wide range of telecommunications products. Trillium's solutions consisted of software products and services for data, voice, and high availability. Lawrence, along with his team, built Trillium from two people, zero products, zero revenue, and $1,000 initially invested capital at its founding in February 1988 to more than 250 people, more than 100 products consisting of millions of lines of source code, $30 million annual revenue, and acquisition by Intel in August 2000 for $300 million. Jeff led, built, motivated, and grew a team having a unique and deep understanding of the communications software industry and having a culture and commitment to the customer and success during a period of continual and rapid change in the technology, market, competitive, and financial environment.

Lawrence is also Trustee of The Lawrence Foundation (www.thelawrencefoundation.org), a family foundation focused on grant making to environmental, education, health, and human services causes. He has a B.S.E.E. from UCLA, was co-recipient of the Greater Los Angeles Entrepreneur of the Year award, and recipient of the UCLA School of Engineering's Professional Achievement award. Jeff also writes and speaks on technology, entrepreneurship, philanthropy, economics, and ethics.

Larry Bock

Larry Bock is the Founder and Executive Chairman of Nanosys Inc. He is a General Partner of CW Ventures, a life sciences venture capital fund, and a Special Advisor to Lux Capital, a nanotechnology-focused venture capital fund. He is a member of the board of directors of FEI Corporation, the leading supplier of tools for nanotechnology research.

Bock was the Founder and initial CEO of Neurocrine Biosciences, Pharmacopeia, GenPharm International, Caliper Technologies, Illumina Technologies, IDUN Pharmaceuticals, Metra Biosystems, and FASTTRACK Systems. Bock was a co-founder of Argonaut Technologies, ARIAD Pharmaceuticals, Athena Neurosciences, Vertex Pharmaceuticals, and Onyx Pharmaceuticals. He also helped found and is on the scientific advisory board of Conforma Therapeutics.

Bock started his career as a Researcher at Genentech, Inc., in the field of infectious diseases, where he was on the team that received the AAAS Newcomb Cleveland Prize for demonstrating the first recombinant DNA vaccine.

He has received several awards and honors. He was selected by *Venture Capital Journal* as one of the "Ten Most Influential Venture Capitalists," by *Red Herring* as one of the "Top Ten Innovators," by *Forbes-Wolfe NanoReport* as the "Number One Powerbroker in Nanotechnology," by Ernst & Young as a finalist for "Entrepreneur of the Year," and by *Small Times* as "Innovator of the Year" and one of "Top 3 Business Leaders of the Year." He received the Einstein Award from the Jerusalem Foundation for lifetime contributions in the field of life sciences.

Bock and his wife, Diane, founded Community Cousins, a nonprofit foundation focused on breaking down racial barriers, which was selected by former Vice President Al Gore as one of ten outstanding grassroots efforts nationally. Bock received his B.A. in biochemistry from Bowdoin College and his M.B.A. from the Anderson School at UCLA.

Jim Duncan

Jim Duncan is the Group Director of Business Development for Meggitt Aerospace Equipment (MAE) Group, a strategic business unit of Meggitt PLC. MAE companies specialize in products and services around the following technologies: thermal management systems, fluid dynamics, engineered materials, extreme environment cabling, and sensors and subsystems.

Duncan is a Partner and Co-founder of Blue Sky Management and Consultants, Ltd. Blue Sky was organized to develop investment, technology transfer, strategic business development, and manufacturing opportunities for technology-oriented companies. Duncan was the Executive Vice President for Business Development of Zone5 Wireless Systems Inc., a Rockwell Science Center (RSC) spin-off company. He was a co-founder of the venture-capital-backed start-up business for wireless ad hoc data network products.

Prior to joining Zone5, Duncan was the Director, Business Development & Commercialization at Rockwell Science Center (RSC). He led the commercialization of RSC's wireless integrated network sensor technology—the technology licensed to Zone5. The product, HIDRA, won industry recognition, receiving Product of the Year awards in 2000 and 2001. The technology addressed the convergence of processing, radio frequency, ad hoc networks, and sensor technologies.

Before joining the Science Center, Duncan co-founded and directed an international (operations on Taiwan and North America) start-up company that designed, produced, marketed, and distributed wireless consumer products—stereo speakers and audio/video surveillance systems. Duncan has held major business development and marketing positions with technology-based companies over the past 20 years: Research (Rockwell Scientific), Antenna and Radar Subsystems (Teledyne Ryan Electronics), Systems Integration (Teledyne Electronic Systems), Aircraft Prime manufacturer (Kaman), Consumer electronics (Paradox), Commercial (Zone5), and consulting (Blue Sky).

Duncan is a former U.S. Naval aviator. While a commissioned officer he flew missions in the Kaman H-2 and Sikorsky SH-60B antisubmarine warfare helicopters. He graduated from the U.S. Naval Academy with a B.S. in physical science.

Meyya Meyyappan

Meyya Meyyappan is Director of the Center for Nanotechnology as well as Senior Scientist at NASA Ames Research Center in Moffett Field, CA (http://www.ipt.arc.nasa.gov). He is one of the founding members of the Interagency Working Group on Nanotechnology (IWGN) established by the Office of Science and Technology Policy (OSTP). The IWGN is responsible for putting together the National Nanotechnology Initiative.

Meyyappan is a Fellow of the Institute of Electrical and Electronics Engineers (IEEE). He is Fellow of the Electrochemical Society (ECS). In addition, he is a member of American Society of Mechanical Engineers (ASME), Materials Research Society, American Vacuum Society, and American Institute of Chemical Engineers. He is the IEEE Distinguished Lecturer on Nanotechnology and ASME's Distinguished Lecturer on Nanotechnology. He is the President-elect of the IEEE's Nanotechnology Council.

For his work and leadership in nanotechnology, he was awarded NASA's Outstanding Leadership Medal and received the Arthur Flemming Award from the Arthur Flemming Foundation and George Washington University. For his contributions to nanotechnology education and training, he received the 2003–2004 Engineer of the Year award from the San Francisco section of the AIAA (American Institute of Aeronautics and Astronautics). In 2004, he was honored with the President's Meritorious Award for his contributions to nanotechnology.

Mark Reed

Mark A. Reed received his Ph.D. in physics from Syracuse University in 1983, after which he joined Texas Instruments, where he co-founded the nanoelectronics research program. In 1990 Reed left TI to join the faculty at Yale University, where he presently holds a joint appointment as Professor in the Electrical Engineering and Applied Physics departments, and is the Harold

Hodgkinson Chair of Engineering and Applied Science. His research activities have included the investigation of electronic transport in nanoscale and mesoscopic systems, artificially structured materials and devices, and molecular-scale electronic transport. Reed is the author of more than 150 professional publications and 6 books, has given 15 plenary and more than 240 invited talks, and holds 24 U.S. and foreign patents on quantum effect, heterojunction, and molecular devices. He has been elected to the Connecticut Academy of Science and Engineering and Who's Who in the World. His awards include *Fortune* magazine's "Most Promising Young Scientist" (1990), the Kilby Young Innovator Award (1994), the DARPA ULTRA Most Significant Achievement Award (1997), the Syracuse University Distinguished Alumni award (2000), the Fujitsu ISCS Quantum Device Award (2001), and the Yale Science and Engineering Association Award for Advancement of Basic and Applied Science (2002). In 2003 he was elected a fellow of the American Physical Society.

Sheryl Ehrman

Sheryl Ehrman received her B.S. in chemical engineering from the University of California, Santa Barbara, in 1991, and her Ph.D. in chemical engineering from the University of California, Los Angeles, in 1997. She has worked as a Visiting Scientist in the Process Measurements Division of the National Institute of Standards and Technology site in Gaithersburg, Maryland. She has also worked as a Post Doctoral Fellow in the Laboratory for Radio and Environmental Chemistry at the Paul Scherrer Institute in Villigen, Switzerland, studying the interactions between ambient particulate matter and ice crystals.

She currently is employed as an associate professor in the Department of Chemical and Biomolecular Engineering and as a participating faculty in both the Bioengineering and the Chemical Physics Programs at the University of Maryland, College Park. Her research interests include the study of nanoparticle formation and processing and bioengineering applications of nanoparticles, as well as the transport and fate of particulate air pollution. She teaches courses in particle technology, transport phenomena, thermodynamics, and introductory engineering design.

Brent Segal

Brent Segal is a member of Echelon Ventures of Burlington, Massachusetts, where he focuses on investments involving combinations of disciplines leading to explosive growth potential. Some of his primary areas of expertise include chemistry, biochemistry, biology, semiconductors, and nanotechnology. He is a Co-founder and part-time Chief Operating Officer of Nantero, a leading nanotechnology company, where he oversees operations roles focusing on partnerships, involving companies such as LSI Logic, BAE Systems, and ASM Lithography. He continues to assist Nantero with intellectual property management and government programs involving the Navy and various agencies.

Segal is an active member of the steering committee of the Massachusetts Nanotechnology Initiative (MNI), executive member of the Massachusetts NanoExchange (MNE), a member of the New England Nanomanufacturing Center for Enabling Tools (NENCET) Industrial Advisory Board, and a member of the planning board for Nanotech 2006. He sits on the board of directors of Coretomic, of Burlington, Vermont, and ENS Biopolymer, Inc., of Cambridge, Massachusetts.

He is an active fund-raiser, helping Nantero to raise more than $36 million in three private equity rounds and government programs. He previously ran laboratory operations at Metaprobe LLC, where he also assisted in raising more than $5 million in private equity financing. He was a Research Associate at Nycomed Salutar, Inc., where he secured several new patents involving novel X-ray contrast agents for medical imaging.

He received his Ph.D. in chemistry from Harvard University in 2000 and has published frequently in journals, including *Journal of the American Chemical Society, Inorganic Chemistry,* and various IEEE publications, including one in which Nantero was named one of the top ten companies for the next ten years. Segal is co-author of more than 50 patents and has worked extensively on intellectual property creation and protection issues at both Nycomed and Metaprobe. He is a graduate of Reed College, with a degree in biochemistry.

Zhong Lin (ZL) Wang

Zhong Lin (ZL) Wang is a Regents' Professor and Director, Center for Nanostructure Characterization and Fabrication, at Georgia Tech. He has authored and co-authored 4 scientific reference books and textbooks, more than 400 peer-reviewed journal articles, and 45 review papers and book chapters; has edited and co-edited 14 books on nanotechnology; and has held 12 patents and provisional patents. Wang is among the world's top 25 most cited authors in nanotechnology from 1992 to 2002 (ISI, Science Watch). He contributed 38 percent of the total citations of Georgia Tech papers in nanotechnology in the past decade, which places Tech number 12 worldwide. His publications have been cited more than 9,000 times.

Wang discovered the nanobelt in 2001, which is considered to be a groundbreaking work. His paper on nanobelts was the second most cited paper in chemistry in 2001 and 2003 worldwide. His paper on piezoelectric nanosprings was one of the most cited papers in materials science in 2004 worldwide. In 1999, he and his colleagues discovered the world's smallest balance, nanobalance, which was selected as the breakthrough in nanotechnology by the American Physical Society. He was elected to the European Academy of Science (www.eurasc.org) in 2002 and made a fellow of the World Innovation Foundation (www.thewif.org.uk) in 2004. Wang received the 2001 S.T. Li prize for Outstanding Contribution in Nanoscience and Nanotechnology, the 2000 and 2005 Georgia Tech Outstanding Faculty Research Author Awards, Sigma Xi 2005 sustaining research awards, Sigma Xi 1998 and 2002 best paper awards, the 1999 Burton Medal from Microscopy Society of America, the 1998 U.S. NSF CAREER award, and the 1998 China-NSF Overseas Outstanding Young Scientists Award. A symposium in honor of Wang was organized by L'Institut Universitaire de France (IUF) on May 7, 2003. He is a member of the editorial boards of 12 journals. His most recent research focuses on oxide nanobelts and nanowires, in-situ techniques for nanoscale measurements, self-assembly nanostructures, and fabrication of nanodevices and nanosensors for biomedical applications. Details can be found at http://www.nanoscience.gatech.edu/zlwang.

Fiona Case

Fiona Case has more than 15 years' experience in the industrial applications of polymer and surfactant science. Her introduction to industrial polymer science was in the late 1980s at Courtaulds Research in the U.K., where she worked as part of the team developing Tencel, a new environmentally friendly solvent-spun cellulose fiber. She also investigated the effects of polyacrylonitrile microstructure on carbon fiber performance using early molecular and quantum mechanics computer modeling techniques. Case's involvement as a founding member of the Biosym/Molecular Simulations Inc. (MSI) Polymer Consortium, one of the earliest materials simulation efforts, meant frequent trips to sunny San Diego, California. In 1991 she moved to California to join Biosym. Case spent nine years at Biosym/MSI (now Accelrys). She carried out contract research for some of the top U.S. and European companies. As manager of the MSI training group, she prepared and presented more than 30 workshops worldwide on polymer science and molecular modeling. She also worked in marketing and technical sales support.

In 1999 she was hired into a central research group at Colgate Palmolive. This presented challenges, including materials structure and property prediction for toothpaste, detergent, hard surface care and personal care products, and packaging and fragrance technology. In 2003 Case left Colgate Palmolive to move to beautiful Vermont with her husband, Martin Case, and to found Case Scientific (www.casescientific.com), offering consultancy and contract research in soft nanotechnology, computational chemistry, and polymer and surfactant science. Case is a Chartered Chemist and a member of the Royal Society of Chemistry, the ACS, the American Oil Chemists Society, and the National Association of Science Writers.

David J. Nagel

David Nagel started school at an early age and completed it shortly before turning 40, apparently a slow learner. Unable to decide on what to do, he studied three majors and wound up with a Ph.D. in materials engineering from the University of Maryland. After four years of active duty and 26 years in the Naval Reserve, he became an expired captain in 1990. He worked at the Naval Research Laboratory for 36 years, apparently unable to find a job elsewhere. While there, he had an accident and turned into a division head for 13 years. His careers in the Navy and in R&D were concurrent, not consecutive. For the past six years, he has been a research professor at The George Washington University. This means that he is neither a real doctor nor a real professor. Currently, Nagel, still unable to focus, performs studies on a wide variety of topics. He teaches a course on applications of micromachines, which is his chance to retaliate for suffering inflicted by teachers during his long academic career. His primary products are hot air and marks on otherwise clean papers.

(Editor's meddling: Nagel was Superintendent of the Condensed Matter and Radiation Sciences Division at the Naval Research Laboratory and has written or co-authored more than 150 technical articles, reports, book chapters, and encyclopedia articles primarily on the applications of MEMS and nanotechnology.)

Sharon Smith

Sharon Smith is a Corporate Executive at Lockheed Martin's headquarters in Bethesda, Maryland. As Director, Advanced Technology, she is responsible for research and technology initiatives, including independent research and development projects, university involvement, and various other R&D activities. She is the prior chair of the Lockheed Martin Steering Group on Microsystems/MEMS and is currently the chair of the corporation's Steering Group on Nanotechnology.

Smith has 25 years of experience in management, program management, engineering, and research and development at Eli Lilly and Company, IBM Corporation, Loral, and Lockheed Martin Corporation. She has more than 25 technical publications and has given numerous technical presentations in the United States and Europe. She serves on the Nanotechnology Committee for JCOTS (Joint Committee On Technology and Science) for the state of Virginia and is also a member of the National Academies' National Materials Advisory Board.

Smith has a Ph.D. in analytical chemistry from Indiana University, an M.S. in physical chemistry from Purdue University, and a B.S. in chemistry from Indiana University.

George Thompson

George Thompson is currently in Intel's Technology Strategy Group. Previously, he has managed projects for new product launches and pathfinding, enterprise platform enabling, materials development, and high-volume manufacturing. His main interests in silicon processing were etch process development and diagnosis, process integration, and yield analysis. His past research interests include applied optics, solid-state laser development, solar energy research, precision frequency control, and molecular spectroscopy. He has a Ph.D. in physical chemistry.

Stephen Goodnick

Stephen Goodnick received his Ph.D. in electrical engineering from Colorado State University in 1983. He is currently Interim Deputy Dean and Director of Nanotechnology for the Ira A. Fulton School of Engineering at Arizona State University. He joined ASU in fall 1996 as Department Chair of Electrical Engineering. Prior to that, he was professor of electrical and computer engineering at Oregon State University from 1986 to 1996. He has also been a visiting scientist at the Solar Energy Research Institute and Sandia National Laboratories and a visiting faculty member at the Walter Schottky Institute, Munich, Germany; the University of Modena, Italy; the University of Notre Dame; and Osaka University, Japan. He served as President (2003–2004) of the Electrical and Computer Engineering Department Heads Association (ECEDHA), and as Program Chair of the Fourth IEEE Conference on Nanotechnology in 2004.

Goodnick has published more than 150 refereed journal articles, books, and book chapters related to transport in semiconductor devices, computational electronics, quantum and nanostructured devices and device technology, and high-frequency and optical devices. He is co-author of the textbook *Transport in Nanostructures* with David K. Ferry (Cambridge University Press, 1997). He is a Fellow of the IEEE (2004), and an Alexander von Humboldt Research Fellow in Germany.

Axel Scherer

Axel Scherer is Professor of Electrical Engineering, Applied Physics, and Physics at the California Institute of Technology, specializing in device microfabrication and packaging. He graduated from New Mexico Institute of Mining and Technology in 1985 and worked in the Quantum Device Fabrication group at Bellcore for the following eight years. In the past, Scherer specialized in improving the state of the art of semiconductor microfabrication, which resulted in the development of the smallest vertical cavity lasers (400nm wide) and some of the world's smallest etched structures (6nm wide), as well as ultranarrow gratings (30nm pitch). He has also been working on reducing the sizes of microlasers and changing their emission wavelengths. Scherer's research laboratory is built around producing nanostructures and applying them to new optoelectonic, magnetooptic, and high-speed electronic devices. The aim of his research group is to develop functional devices that use their reduced geometries to obtain higher speed and greater efficiencies, and can be integrated into systems in large numbers.

Suzie Hwang Pun

Suzie Pun is an assistant professor in the Department of Bioengineering at the University of Washington. Her research focuses on synthetic gene and drug delivery systems. She received her B.S. in chemical engineering at Stanford University. For her graduate work, she developed polymeric materials for drug delivery and obtained her Ph.D. in chemical engineering from the California Institute of Technology under the guidance of Mark E. Davis in 2000. She then worked at Insert Therapeutics as a Senior Scientist, pioneering the use of cyclodextrin-containing polymer delivery systems for nucleic acids. She joined the faculty at University of Washington in 2003 and has published 16 papers in the field of drug delivery. In addition, she has 8 granted or pending patent applications.

Pun has been recognized by *MIT Technology Review* magazine as one of 2002's "Top 100 Young Innovators" in its TR100 Awards. She is the recipient of a Career Development Award from the National Hemophilia Foundation

(2004), a Young Investigator Award from the Alliance for Cancer Gene Therapy (2005), and a Faculty Early Career Development (CAREER) award from the NSF (2005).

Jianjun (JJ) Cheng

Jianjun Cheng is an Assistant Professor in the Department of Materials Science and Engineering of the University of Illinois at Urbana-Champaign. Cheng received his B.S. in chemistry at Nankai University and his M.S. in chemistry from Southern Illinois University. He then received his Ph.D. in materials science from the University of California, Santa Barbara. His industry experience includes working as a Senior Scientist and Project Leader for Insert Therapeutics, Inc., where he developed Cyclodextrin-based polymers for systemic delivery of camptothecin and the modification of polycations for systemic gene delivery and targeting. He has also done postdoctoral research at the Massachusetts Institute of Technology with Robert Langer, an MIT Institute professor who has written more than 840 articles, has more than 500 issued or pending patents, and has received the Charles Stark Draper Prize, considered the equivalent of the Nobel Prize for engineers. In this position Cheng researched magnetic nanoparticles for oral protein delivery and Aptamer-nanoparticles for in vivo targeted prostate cancer therapy and diagnosis. Cheng has more than 20 peer-reviewed publications and patent applications.

Dan Garcia

Dan Garcia is currently a Ph.D. student in the Department of Bioengineering at UCLA. He works in the laboratory of Dr. Chih-Ming Ho, where he is using NEMS and nanofluidic technology to develop the actin muscle protein for a bottom-up/top-down hybrid nanofabrication process to construct two- and three-dimensional nanostructures. This will ultimately facilitate the generation of actin filaments according to a two- or three-dimensional, predesigned layout.

Prior to joining Dr. Ho's group, Garcia conducted immunology research in the laboratory of Dr. Genhong Cheng. Garcia's work helped clarify the interaction between a novel protein, TRAF3-interacting JunN-terminal kinase (JNK)-activating modulator (T3JAM), and tumor necrosis factor receptor (TNFR)-associated factor 3 (TRAF3), two proteins that play important roles in the CD40 signal transduction pathway. The CD40 signaling pathway plays a role in the adaptive immune response, atherosclerosis, and the body's response to transplants.

Dean Ho

Dean Ho received his B.S. in physiological science from UCLA in 2001, and his Ph.D. in 2005 from the UCLA Department of Bioengineering. He is currently a Hewlett-Packard Fellow in the Department of Mechanical and Aerospace Engineering at UCLA. His research has covered emerging areas in bionanotechnology to interface membrane-bound proteins with block copolymeric biomimetic membranes to fabricate biomolecular hybrids. Ho has been among the first to demonstrate the coupling of protein function and the potential of using biostructures as energy conversion systems. This work is the subject of a featured article in *Nanotechnology* that was downloaded more than 1,000 times, which was among the top 10 percent of all downloaded Institute of Physics (IOP) publications, as well as a feature article in *Advanced Functional Materials.* He has published more than 20 peer-reviewed papers and several book chapters in the areas of biochemical energetics and biotic-abiotic interfacing, as well as developing biologically active devices based on the harnessing of protein functionality. Ho's research achievements have garnered news coverage in *Nature, MICRO/NANO,* and BBC Radio, where Ho was interviewed on the subject of the societal impacts of bionanotechnology. He has presented several invited talks at internationally renowned institutions, including the California Nanosystems Institute (CNSI), Academia Sinica, and the Bionanotechnology: Academic and Industrial Partnerships in the U.K. and U.S. conference held at the University of Southern California. Ho is a member of Sigma Xi, the Biomedical Engineering Society, Materials Research Society, IEEE, AAAS, ASME, and American Academy of Nanomedicine. In addition, he currently serves on the editorial board of the *Journal of Nanotechnology Law and Business* and previously served as the Editor-in-Chief of the *CESASC Technical Symposium Proceedings* and as an Editor of the UCLA *Graduate Scientific Review.*

In addition to his academic accomplishments, Ho has made extensive contributions to the UCLA community, having previously served as a Trustee-at-Large of the Unicamp Board of Trustees, the body that oversees the official philanthropy of UCLA that is currently in its 70th year of activity. For his service and leadership to the UCLA and greater Los Angeles community, Ho was presented with the UCLA Chancellor's Service Award in 2001.

Chih-Ming Ho

Chih-Ming Ho is the Ben Rich-Lockheed Martin Chair Professor at the UCLA School of Engineering and the Director of the Institute for Cell Mimetic Space Exploration (CMISE). He graduated from the Mechanical

Engineering Department of the National Taiwan University. After receiving his Ph.D. from The Johns Hopkins University, Ho started his career at the University of Southern California and rose to the rank of Full Professor. In 1991, he moved to the University of California, Los Angeles, to lead the establishment of the microelectromechanical system (MEMS) field in UCLA and served as the founding Director of the Center for Micro Systems. The UCLA MEMS research has been recognized as one of the best in the world. He served as UCLA Associate Vice Chancellor for Research from 2001 to 2005.

He is an internationally renowned scientist in micro- and nanofluidics, bio-nano technology, and turbulence. He was ranked by ISI as one of the top 250 most cited researchers in all engineering categories around the world, and by ECI as number 7 in the field of MEMS. In 1997, Ho was inducted as a member of the National Academy of Engineering. In the next year, he was elected as an Academician of Academia Sinica, which honors scholars of Chinese origin with exceptional achievements in liberal arts and the sciences. Ho holds five honorary professorships. He has published 220 papers and 7 patents. He has presented more than 100 keynote talks in international conferences. Ho was elected fellow of the American Physical Society as well as American Institute of Aeronautics and Astronautics for his contributions in a wide spectrum of technical areas.

In addition to his academic accomplishments, he has made extensive contributions to professional societies around the world. He has chaired the Division of Fluid Dynamics (DFD) of the American Physical Society, which is the leading platform in the United States for scientists interested in fundamental fluid dynamics. He is on the advisory board for *AIAA Journal* and is a member of the IEEE/ASME JMEMS coordinating committee. He was an Associate Editor of the *ASME Journal of Fluids Engineering* and an Associate Editor of the *AIAA Journal.* He also has served as a Guest Editor for *Annual Review of Fluid Dynamics.*

On the international level, Ho has served on advisory panels to provide assistance to many countries and regions—France, China, United Kingdom, Israel, Taiwan, and Japan—on the development of nano- and microtechnologies. Ho also has chaired or served on numerous organizing committees of international conferences on high-technology topics.

Mihail C. Roco

Mihail Roco is Co-Chair of the U.S. National Science and Technology Council's Subcommittee on Nanoscale Science, Engineering and Technology (NSET), and is Senior Advisor for Nanotechnology at the National Science

Foundation. He also coordinates NSF's Grant Opportunities for Liaison with Industry program. Previously he was Professor of Mechanical Engineering at the University of Kentucky (1981–1995) and held visiting professorships at the California Institute of Technology (1988–1989), Johns Hopkins University (1993–1995), Tohoku University (1989), and Delft University of Technology (1997–1998).

Roco is credited with 13 inventions. He has authored or co-authored numerous archival articles and twelve books, including *Particulate Two-phase Flow* (Butterworth, 1993), *Nanostructure Science and Technology* (Kluwer Acad., 1999), *Societal Implications of Nanoscience and Nanotechnology* (Kluwer Acad., 2001), *Converging Technologies for Improving Human Performance* (Kluwer Acad., 2003) and *The Coevolution of Human Potential and Converging Technologies* (N.Y. Acad. of Sciences, 2004). Roco was a researcher in multiphase systems, visualization techniques, computer simulations, nanoparticles, and nanosystems in 1980s as Professor, and in 1991 initiated the first federal government program that focused on nanoscale science and engineering (on Synthesis and Processing of Nanoparticles at NSF). He formally proposed NNI in a presentation at White House/OSTP, Committee on Technology, on March 11, 1999. Roco is a key architect of the National Nanotechnology Initiative and coordinated the preparation of the U.S. National Science and Technology Council reports "Nanotechnology Research Directions" (NSTC, 1999) and "National Nanotechnology Initiative" (NSTC, 2000).

Roco is a Correspondent Member of the Swiss Academy of Engineering Sciences, a Fellow of the American Society of Mechanical Engineers, a Fellow of the Institute of Physics, and a Fellow of the American Institute of Chemical Engineers. He is Editor-in-chief of the *Journal of Nanoparticle Research* and has served as Editor for the *Journal of Fluids Engineering* and the *Journal of Measurement Science and Technology.* He is a member of the Executive Governance Board for Sandia and Los Alamos National Laboratories, the Review Board for National Research Council Institute (Canada), International Risk Governance Council, and other boards in Europe and Asia.

Roco was selected as "Engineer of the Year" by NSF and the U.S. National Society of Professional Engineers in 1999 and again in 2004. Among his other honors are Germany's Carl Duisberg Award, a Burgers Professorship Award in Netherlands, the U.S. University Research Professorship award, and a 2002 Best of Small Tech Award as "Leader of the American Nanotechnology Revolution." *Forbes* magazine recognized him in 2003 as first among "Nanotechnology's Power Brokers," and *Scientific American* named him one of 2004's "Top 50 Technology Leaders."

William Sims Bainbridge

William Sims Bainbridge earned his doctorate in sociology from Harvard University. He is the author of 11 books, 4 textbook-software packages, and about 180 shorter publications in information science, social science of technology, and the sociology of religion. Most recently, he is the Editor of the *Berkshire Encyclopedia of Human-Computer Interaction* and author of the forthcoming *God from the Machine,* a study using artificial intelligence techniques to understand religious belief. With Mihail Roco, he co-edited *Societal Implications of Nanoscience and Nanotechnology* (2001), *Societal Implications of Nanoscience and Nanotechnology II* (2005), *Converging Technologies for Improving Human Performance* (2003), and *Converging Technologies in Society* (2005). At the National Science Foundation since 1992, Bainbridge represented the social and behavioral sciences on five advanced technology initiatives—High Performance Computing and Communications, Knowledge and Distributed Intelligence, Digital Libraries, Information Technology Research, and Nanotechnology—before joining the staff of the Directorate for Computer and Information Science and Engineering. More recently he has represented computer science on the Nanotechnology initiative and the Human and Social Dynamics initiative. Currently, he is Program Director for Social Informatics, after having directed the Sociology, Human Computer Interaction, Science and Engineering Informatics, and Artificial Intelligence programs.

SECTION ONE

Development Drivers

Chapter 1

Lessons in Innovation and Commercialization from the Biotechnology Revolution

Gerald Gallwas

"We shape our tools and forever after, they shape us."
—Marshall McLuhan

Chemistry in many forms is a major force shaping the modern world. It is the fundamental science of materials, providing everything from the tools we use to the medicines we take. Silicon and germanium enabled integrated circuits, hydrogen and oxygen propelled us into space, and recombinant DNA brought new drugs and pest-resistant crops.

Yet for all these benefits, chemistry is both a misunderstood, and in many ways an unappreciated, science. Rather than praise its benefits, our attention is repeatedly focused on chemistry's perils, such as its potential to pollute. Nevertheless, we have at our disposal a great abundance of chemical technology that is changing our lives without our even knowing it. An ancient Chinese proverb reminds us that "only a fool would predict the future." As we contemplate that future, the wiser question to ask is, what did we learn from chemistry in the twentieth century? The story of biotechnology offers many contemporary lessons, especially in innovation and the commercialization of new technologies.

THE STORY OF BIOTECHNOLOGY

Although the story of biotechnology began much earlier, it was Jim Watson and Francis Crick's 1953 announcement of the structure of DNA that made the new biology a public issue.[1] The double-helix model they described was simple and eloquent. It provided an explanation for the replication of life at the molecular level.

An avalanche of research followed, but it was not until 1972 that Paul Berg and colleagues at Stanford University developed techniques to "cut and paste" DNA, creating the first recombinant DNA molecule containing DNA from two different species.[2] As an analogy, think of editing an audiotape. Suppose you cut the tape with a pair of scissors at a point between two songs, insert a new piece of tape with a new song at the cut, and use an adhesive to join the ends together. If you insert the spliced tape into a tape recorder, you'll hear a different sequence of songs.

Even at this early stage, Berg postulated that recombinant DNA technology could provide a pathway to gene therapy. Berg shared the 1980 Nobel Prize in chemistry for this pioneering work. In 1973, Herbert Boyer (of the University of California, San Francisco) and Stanley Cohen (of Stanford University) were credited with being the first "genetic engineers"; they used restriction enzymes to selectively cut and paste DNA and then inserted the new DNA into bacteria, which, by reproducing, made millions of copies of the new DNA. In effect, this was the creation of a DNA factory.[3]

The news spread. Advances in recombinant DNA were exciting and yet frightening. Scientists, then the news media, and finally Congress became concerned about the implications of gene manipulation. Would it create a modern Frankenstein's monster? What controls should be imposed? In early 1975, more than 100 interested parties from around the world, guided by the leadership of Paul Berg, met at the Asilomar Conference Center in California to debate the promise and pitfalls of recombinant DNA.[4] The recommendations of the meeting were sent to the National Institutes of Health (NIH) and served as the basis for the official guidelines for the technology, published in 1976.[5]

Boyer and Cohen's success in inserting a gene into bacteria aroused the interest of a young venture capitalist, Robert Swanson, from San Francisco.[6] In 1976, Swanson's question was, how could the Boyer-Cohen technology be used to produce a marketable protein product, perhaps a therapeutic product such as human insulin? In April of that year, Swanson and Boyer each invested $500 to create Genentech, the world's first biotechnology company. Competition quickly followed, with the formation of Biogen. The flag of the biotech industry had been planted, and the first goal was to create human insulin.

Genentech and Biogen chose different technical paths to this common goal. Genentech focused on chemically synthesizing the human insulin gene, whereas Biogen relied on cloning techniques. The chemically synthesized gene would escape the NIH safety regulations, but the cloning of the human gene would be subject to the NIH controls.

At the time, Genentech was a company in name only. It had no staff, no facilities, and no money. Boyer turned to Arthur Riggs and Keiichi Itakura—two colleagues at the City of Hope National Medical Center—and asked whether they would accept a contract to synthesize the human insulin gene. Riggs and Itakura were in the process of writing a grant proposal to the NIH for the synthesis of the human hormone somatostatin—a simpler task, but one that would demonstrate the pathway to the synthesis of insulin.

Riggs answered Boyer's question with a question: Would Genentech be willing to sponsor the somatostatin project first? The answer was yes, and the Genentech and City of Hope teams joined forces. Soon thereafter, Riggs and Itakura inserted a 21-nucleotide-long strand of DNA into *E. coli* bacteria, and with the collaboration of Herb Heyneker, a young chemist from Boyer's lab, they demonstrated for the first time that man-made DNA would function in a living cell.

Sixteen months later, the combined team successfully synthesized a gene for somatostatin, cloned it, and expressed the somatostatin protein hormone in bacteria.[7] This was the first time anyone had successfully expressed a protein of any kind in genetically engineered bacteria. The race for human insulin accelerated.

The Riggs-Itakura technology was remarkable because it was a general technique, one that could be used to manufacture most proteins in a bacterial host. It resulted in numerous U.S. and foreign patents, from which billions of dollars' worth of pharmaceutical products have been developed. In the meantime, the NIH turned down Riggs and Itakura's grant proposal on the grounds that it was too ambitious and without practical purpose!

With the somatostatin success, Swanson began to raise funds in earnest, holding out the real promise of human insulin to investors. By June 1978, Genentech was hiring scientists and had built a laboratory facility near the San Francisco airport. Three months later, near the end of August, the combined City of Hope and Genentech team had produced human insulin from a synthesized gene. The improbable had become reality.

The story of Genentech and the launch of the biotech industry is well told in histories by Hall and by Evans.[8, 9] It is the story of basic science from an academic environment turned into a phenomenal business success. It is the story of an industry starting from scratch, not a spin-off from a mature business sector, as with pharmaceuticals.

Today, nearly thirty years later, biotechnology is on its way to being a trillion-dollar industry producing hundreds of therapeutic and biological products.[10] As we anticipate the future development of nanotechnology, lessons from the biotechnology revolution can offer a useful perspective, particularly with respect to two concepts fundamental to an understanding of technology innovation and commercialization.

CONCEPT 1: LESSONS FROM THE S-CURVE

The first concept is that of the S-curve. Introduced by Richard Foster, the S-curve describes the relationship between the investment of resources in a technology or process and the resulting performance produced by that investment (Figure 1–1).[11] It is referred to as an S-curve because when the results are plotted, the resulting figure typically takes the shape of the letter *S*.

In an early stage, as effort is ramping up on a technology or process, there is little return on investment while basic knowledge is slowly gained. As the knowledge base increases, the investment passes a tipping point; after that point, progress accelerates rapidly and then begins to slow as the technology reaches maturity. Beyond a certain point, no amount of additional investment will markedly improve the performance.

The single-engine propeller aircraft serves as an excellent example (Figure 1–2).[12] The Wright brothers' first flight in 1903 was at a speed of about 35 mph. Seven years later, in 1910, the Gordon Bennett air race at Belmont Park,

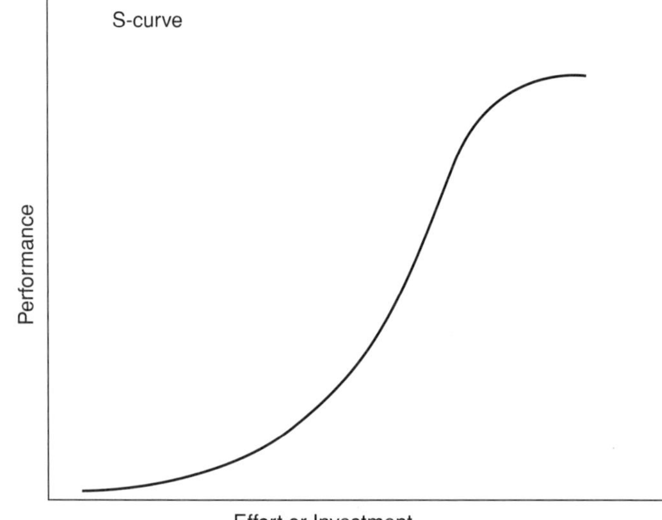

Figure 1–1 An example of an S-curve.

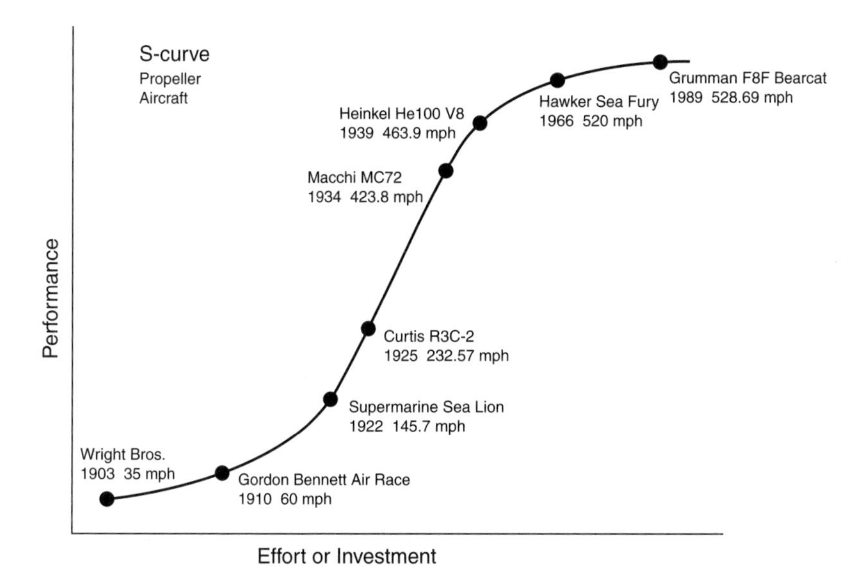

Figure 1–2 S-curve, single-engine propeller aircraft example.

New York, was won at 60 mph. As the fundamentals of aircraft design became widely understood, speeds increased to 139.66 mph in 1914 and then to 145.7 mph in 1922. In 1925, Lieutenant James Doolittle, in a Curtiss R3C-2, flew at 232.57 mph. By 1939, the record stood at 463.9 mph, and the S-curve began to flatten. In 1966, some 27 years later, a new record for a single-engine piston aircraft was set at the Reno (Nevada) Air Races at 520 mph. By 1989, that number had improved by less than 9 mph. Somewhere in the vicinity of 525 mph, the limits of single-engine propeller technology were reached. No amount of investment will return significantly more performance.

Jet aircraft engine technology offered the next step in performance and, as you might suspect, defined a new S-curve in air speed. The gap between the two curves is referred to as the *discontinuity* between the two technologies (Figure 1–3).

The example of the single-engine propeller-driven aircraft offers two lessons that are characteristic of innovation and commercialization: the lessons of limits and of discontinuity. Low on the curve there is high risk, great uncertainty, and high potential for growth. High on the curve there is performance certainty, giving rise to a false sense of security and yet little opportunity for growth. The only pathway to significantly greater performance is to jump to a new S-curve—from propeller to jet, from vacuum tube to transistor, from the horse-drawn vehicle to the automobile. These are the lessons of limits and discontinuities.

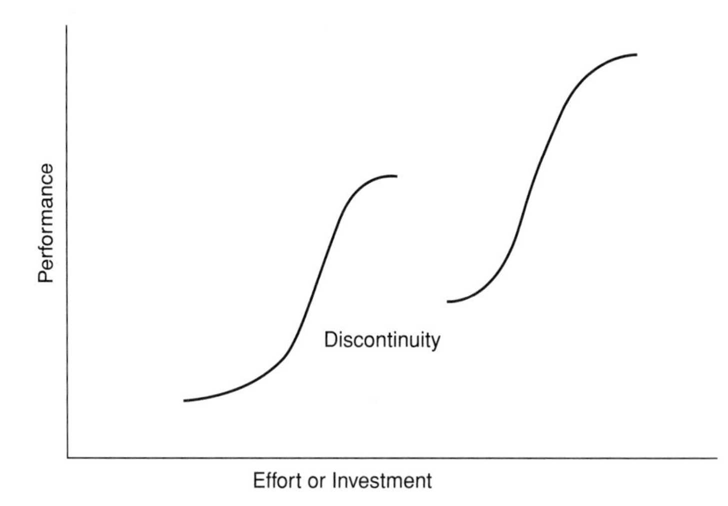

Figure 1–3 S-curves often come in pairs.

In the case of biotechnology, the original S-curve might be characterized as the technology for commercialization of animal proteins for human use, as in the case of insulin. Bovine and porcine insulin were extracted from slaughterhouse animals and then purified and sold. The discontinuity created by the second S-curve involved the use of the synthesized insulin gene inserted into bacteria for the production of human insulin. This innovation provided the basis for an entirely new industry.

CONCEPT 2: LESSONS FROM COMMERCIAL INNOVATION

The second concept that is useful in understanding the commercialization of innovative technology relates to the basic drivers of successful new product or process innovation. Simplistically, innovation has been described as a function of creativity and implementation. But it is more than that. The biotechnology experience, among other examples, shows that innovation should be viewed more comprehensively as a function of

- Creativity
- Management
- Team
- Investment
- Demand

It required creative people with innovative ideas, strong, focused management, a multidisciplinary team, significant investment of capital, and the promise of return on investment to bring biotechnology from the research laboratory to the marketplace.

Innovative Product Commercialization, a function of:

(Creativity × Management × Team)(Investment)(Demand)

As an analogy, consider a musical concert. Creativity is contributed by the composer, management by the music director, and the team by the performing artists and members of the orchestra. Investment in the concert is a matter of the sponsor's belief that the musical experience will attract attendance (demand) and that box office receipts will supply the expected revenue and return on investment. All are required for success.

In the insulin example, the Boyer-Cohen technology clearly provided the creative catalyst that inspired Robert Swanson to lead and manage the commercial development of chemically synthesized insulin. But without the combined team at the City of Hope and Genentech, the race for genetically engineered insulin would not have been a West Coast victory. Swanson's ability to both manage the start-up of Genentech and raise investment capital based on the promise (demand attraction) of insulin was remarkable. But it was the combined multifunctional team of Genentech and the City of Hope, with leverage from the Riggs-Itakura technology, that turned the game entirely in the favor of the Genentech success.

THE S-CURVE SUMMARIZED

Here is the concept of the S-curve in a nutshell:

- New technology is a slow process, requiring years of incubation. Often, new technologies appear to be solutions looking for problems. Some 22 years elapsed between the invention of powered flight and Doolittle's flight at 232 mph. Similarly, 24 years elapsed between Watson and Crick's DNA announcement in 1953 and development of the general technique to manufacture proteins in bacterial hosts based on the somatostatin model created by Riggs and Itakura in 1977. Typically, 20 or more years are required for a new technology to mature to the point of an accelerating S-curve. The knowledge base and tools must be put into place. Eventually, limits are reached; but biotechnology is still relatively low on the S-curve, with hundreds of new products in the development process.

- Every technology has limits. The process of insulin production based on animal source materials was inadequate to meet rising future demand.
- When new technology is introduced, there is skepticism from all quarters, including other scientists, the news media, Congress, government agencies, and the investment community. The discontinuity between the two S-curves means leaving the security of the old for the uncertain promise of the new—a required step to reach new heights of performance.

COMMERCIAL INNOVATION SUMMARIZED

The commercial application of innovation can be summarized as follows:

- New technology requires creative thinking and innovative action. The creative discovery or invention is essential, but it is the implementation through strong management and multidisciplinary team performance that enables commercial success.
- The value of new technology often goes unrecognized and unfunded until the potential is put into perspective. The Riggs-Itakura somatostatin grant request to NIH was rejected as too ambitious and impractical. However, with the announcement of the success of somatostatin and the publicity that followed, the scientific community, the news media, and even Congress took notice.
- Early visionary champions drive the diffusion of new technology. In the case of biotechnology, Robert Swanson was the visionary and champion who inspired Boyer and the multibillion-dollar industry that followed.
- To attract capital for products and processes that are low on the S-curve, investor knowledge and education are required. Somatostatin drew little investment interest until its value was clearly demonstrated by way of a process to create human insulin. It was that promise that drove investment.

THE FUTURE OF NANOTECHNOLOGY

Only a fool would predict the future.—Ancient Chinese Proverb

Predicting the future is a hazardous task. What we need is a tool chest of relevant experience and intuition, one that enables us to move from today's paradigm to a vision of tomorrow's potential. These tools are needed because the rules of the physical world change when technology moves to the atomic level. Our everyday understanding of mass and inertia, which govern actions such as driving an automobile, no longer apply at the nanoscale.

For example, think of inserting a straw into a glass of water. Based on what we have been taught in the macro world, water seeks its own level. Thus, we would expect the water inside the straw to remain at the same level as the water on the outside of the straw, and that is the case. Yet if you were to dip a capillary tube into the same glass of water, the water would rapidly rise in the capillary far above the level of water in the glass. The rules change as we cross from our world of macro reality into the dimension of the capillary, and so it will be with nanotechnology.

Only scientists armed with a sound understanding of the fundamental principles are likely to explore, discover, and invent. They will be questioned and challenged every step of the way. Such was the case with Riggs, Itakura, and Boyer in the example of synthetic insulin. Somatostatin was an experiment of high potential and promise to the City of Hope scientists, and yet even the expert reviewers at the National Institutes of Health failed to recognize the value of the Riggs-Itakura grant proposal. Boyer, well versed in the science, understood the logic and so was able to sell the idea to Swanson. This secured financial support for the somatostatin experiment based on the promise that it provided the pathway to insulin. And so it did.

In the brief history and concepts offered in this chapter, a greatly simplified view has been taken. Fundamental to any view of technological innovation is the discontinuity between two S-curves—in this case, the step from biotechnology to nanotechnology. In the biotechnology example, we relied on nature—enzymes—to do the molecular manipulation of cutting and pasting DNA molecules, and we relied on bacteria to provide the factories for creation of new sources of materials. As we move into the nanotechnology era, all the experience and tools of chemistry and biology in general and recombinant DNA in particular are at our disposal. But rather than rely on enzymes and bacteria, we must now create new nanoscale tools and processes, and it will take time. The ensuing commercialization pathway requires a number of well-defined elements. After discovery come the requirements of demand, investment, and implementation.

By way of example, William Coblentz of the National Bureau of Standards in Washington, D.C., first examined and documented the relationship between the chemical structure of molecules and infrared absorption spectra in 1905.[13] Yet even with the substantial knowledge base that grew from this original work, the first commercial infrared spectrophotometer (an instrument that measures light absorption as a function of wavelength) was not produced until 1942, when it was driven by demand: the urgent need during World War II to produce synthetic rubber. Commercial progress follows as technology evolves from the science to fulfill unmet needs in the marketplace.

In the case of nanotechnology it will be evolutionary and ubiquitous, and not revolutionary and spontaneous.

Predicting the future is a hazardous task, but you can try it by conducting this thought experiment. Consider the world before synthetic fibers, such as nylon; then jump back to the earlier S-curve of natural fibers: hemp, silk, and cotton. With your feet planted in that earlier S-curve, use your tool chest of relevant experience and "enlightened" intuition to envision the world of synthetic fibers. Describe their characteristics; envision the ubiquitous applications that these new materials would enable. You have just experienced the S-curve discontinuity. Now consider the drivers required for the new synthetic technology to become commercial reality.

You have just forecast the future of nanotechnology. Read on.

REFERENCES

1. J. D. Watson and F. H. C. Crick, *Nature* 171 (April 25, 1953): 737.
2. D. A. Jackson, R. H. Symons, and P. Berg, *Proc. Nat. Acad. Sci. USA* 69 (1972): 2904–2909.
3. Stanley N. Cohen, Annie C. Y. Chang, Herbert W. Boyer, and Robert B. Helling, "Construction of Biologically Functional Bacterial Plasmids *In Vitro*," *Proceedings of the National Academy of Sciences* (1973).
4. P. Berg et al., "Potential Biohazards of Recombinant DNA Molecules," letter, *Science* 185 (1974): 303.
5. *Federal Register* 41, no. 131 (1976), 27911–27943.
6. S. S. Hughes, *Isis* 92 (2001): 541–575.
7. K. Itakura et al., *Science* 198 (1977): 1056–1063.
8. S. S. Hall, *Invisible Frontiers: The Race to Synthesize a Human Gene* (Oxford: Oxford University Press, 2002)
9. H. Evans, *They Made America* (New York: Little, Brown, 2004).
10. J. D. Watson, *DNA, The Secret of Life* (New York: Alfred A. Knopf, 2004).
11. R. Foster, *Innovation: The Attacker's Advantage* (New York: Summit Books, 1986).
12. M. W. Bowman, *The World's Fastest Aircraft* (Wellingborough, UK: Patrick Stephens Limited, 1990).
13. W. Coblentz, *Investigations of Infrared Spectra* (Washington, DC: National Bureau of Standards, 1905).

Chapter 2

Nanotechnology and Our Energy Challenge

Richard Smalley

I've increasingly begun to believe that a grand challenge before us—the world and humanity—is what I call "the terawatt challenge." I've been looking into the energy issue and I've been on a quest for terawatts, or trillions of watts of electricity. What I'm looking for is to solve what I believe is the most important problem facing humanity, the problem of generating energy of the magnitude that we will need in this century for what could very well turn out to be ten billion people on the planet.

Energy is the single most important challenge facing humanity today. As we peak in oil production and worry about how long natural gas will last, life must go on. Somehow we must find the basis for energy prosperity for the twenty-first century. We should assume that by the middle of this century we will need to at least double world energy production from its current level, with most of this coming from some clean, sustainable, carbon dioxide–free source. For worldwide peace and prosperity, it must be cheap.

Energy is a $3 trillion a year enterprise, by far the biggest enterprise of humankind. The second most important is agriculture. It used to be that agriculture was almost everything, but agriculture is now only half of energy. All of Defense, both in the United States and around the planet, is only $0.7 trillion. I want to find a new oil, an energy source that will do for this century what oil did in the last century.

I have asked audiences across the country, "What do you think deserves to be on the list of world problems?" My hypothesis is that if you gather any group of people together the word *energy* will always appear on this list. I've done this with fourteen independent audiences, and energy, phrased in one way or another, is one of the first problems suggested.

World Problems

1. Energy
2. Water
3. Food
4. Environment
5. Poverty
6. Terrorism and war
7. Disease
8. Education
9. Democracy
10. Population

My second hypothesis is that if you solve the energy problem you will find that at least five of the remaining nine problems on the list now have a path to an answer that's acceptable, whereas in the absence of solving the energy problem, it is difficult, if not impossible in most cases, to have imagined an answer.

Water is a very brutal problem. Either you've got it or you don't. Luckily, on our planet, there's plenty of water. In fact, we probably have more water than anything else. But it has salt in it, and it's often thousands of miles away from where we need it. We need it in vast amounts, hundreds of millions of gallons a day. We can take the salt out of the water. There's no doubt some nanotechnology that will do it at 100 percent efficiency or close to it, or you can just boil the water. You can do it if you find the energy. And you've got to put it in a pipe and pump it to where you need it, which might be thousands of miles away. We can make pipes and we know how to pump them, but it costs energy to make them, it costs energy to maintain them, and energy to pump the water from here to there. We can do it. If we have the energy, we can solve the water problem. Just solve it, for ten billion people everywhere in the planet. If you haven't got the energy, you can't solve the problem.

The third one on the list is food, which is going to be an increasing problem on our planet. We need agriculture and we need water. So if you solve the water problem, you've gone a long way to solving the food problem. In addition to food and water, you need fertilizers. You need energy for that. We need to have a structure where we harvest food and move it around. We need energy for that also. If you've got energy, you can hack this problem. If you don't have cheap, fast energy, I don't see how you can solve the problem.

Virtually every aspect of our fourth problem, the environment, has directly to do with either the way we generate energy, where we get it, how we store it, or what happens when we generate it. I don't know of any other single thing, if you care about the environment, that would have a bigger positive effect than to solve the energy problem. I don't know of anything else you could do for the environment that would be more effective.

Energy would have a tremendous impact if you could solve it—make it cheap, make it abundant, find a new oil. Miracles of science and technology in the physical sciences are primarily what enables this.

My third hypothesis is that no matter what you suggest as being items that deserve to be on this list of the top ten, if you take something other than energy and move it to the top, you will not find anywhere near the same cooperative, positive effect on the other ones as energy.

So you look at where the power is available, the true energy source that's available to generate terawatts. It turns out it's very simple. The big place where we can get the terawatts we need is in the place where there might as well be nothing right now. The technology that enables that has to happen a lot sooner than 2050, because of the huge size of the population. So this is what I call the terawatt challenge: to find a way of changing the energy source so that we can still be prosperous.

It's actually not clear that there is enough coal that's really efficiently, cheaply producible. It's interesting that when you look for terawatts, all the other answers are nuclear. Solar energy is a nuclear energy source, of course. Thus if you like nuclear and if you like nuclear fusion, you'll love the sun. Every day, even though vast amounts of the solar energy go someplace else, 165,000 terawatts hit the Earth. We need only 20 terawatts to completely solve the world's energy needs. This is a vast energy source. We just don't know how to get it cheaply yet.

TRANSPORT AND STORAGE

I've done a lot of thinking about this, trying to think of some scheme for energy around the planet that actually makes sense, technologically and economically, that's up to this terawatt challenge we have. I call this idea the distributive storage grid. In the year 2050, instead of transporting energy by moving masses of coal or oil, you transport energy as energy. You do that with a grid of electrical connections, an interconnected grid of hundreds of millions of local sites.

Consider, for example, a vast interconnected electrical energy grid for the North American continent from above the Arctic Circle to below the Panama Canal. By 2050 this grid will interconnect several hundred million local sites. There are two key aspects of this future grid that will make a huge difference: (1) massive long-distance electrical power transmission and (2) local storage of electrical power with real-time pricing.

Storage of electrical power is critical for the stability and robustness of the electrical grid, and it is essential if we are ever to use solar and wind as our dominant primary power sources. The best place to provide this storage is locally, near the point of use. Imagine that by 2050 every house, every business, every building has its own local electrical storage device, an uninterruptible power supply capable of handling the entire needs of the owner for 24 hours. Because the devices are, ideally, small and relatively inexpensive, the owners can replace them with new models every five years or so as worldwide technological innovation and free enterprise continuously and rapidly develop improvements in this most critical of all aspects of the electrical energy grid.

Today, using lead-acid storage batteries, such a unit for a typical house to store 100 kilowatt hours of electrical energy would take up a small room and cost more than $10,000. Through revolutionary advances in nanotechnology, it may be possible to shrink an equivalent unit to the size of a washing machine and drop the cost to less than $1,000. With these advances the electrical grid can become exceedingly robust, because local storage protects customers from power fluctuations and outages. Most importantly, it permits some or all of the primary electrical power on the grid to come from solar and wind.

The other critical innovation needed is massive electrical power transmission over continental distances, permitting, for example, hundreds of gigawatts of electrical power to be transported from solar farms in New Mexico to markets in New England. Then all primary power producers can compete, with little concern for the actual distance to market. Clean coal plants in Wyoming, stranded gas in Alaska, wind farms in North Dakota, hydroelectric power from northern British Columbia, biomass energy from Mississippi, nuclear power from Hanford, Washington, and solar power from the vast western deserts. Remote power plants from all over the continent contribute power to consumers thousands of miles away on the grid. Nanotechnology in the form of single-walled carbon nanotubes (aka "buckytubes"), forming what we call the "Armchair Quantum Wire," may play a big role in this new electrical transmission system.

Such innovations—in power transmission, power storage, and the massive primary power-generation technologies themselves—can come only from miraculous discoveries in science, together with free enterprise in open competition for huge worldwide markets.

The key is not only an energy source but also energy storage and energy transport. If you can solve the problem of local storage, you've basically solved the whole problem. That's because, by definition, the storage you need is local. You've got terawatts of power moving into the grid. The biggest problem with renewable energies in general, and solar and wind in particular, is that they're episodic and not dispatchable. You've got to have storage. Storing energy in batteries, capacitors, fuel cells, and some chemical systems like hydrogen depends on nanoscale interactions. It depends on charge transfer reactions that take place over the span of a few atoms on their surface. The next generation of storage devices are all optimized by nanoengineered advances and the use of nanoscale catalyst particles. Nanotechnology is where the action is in energy storage.

ENERGY FOR EVERYONE

So in summary I believe that if you had to pick one issue that's more important than any other single issue, it is energy. You could have picked any issue, but energy is something we can do something about, and if we do it, we can affect positively most of the other problems facing us.

The task would be supplying energy for ten billion people, and let's get it clean, cheap, and continually available. If there's an answer, it will be some manipulation of matter on the nanoscale that will give it to us—something I like to call nanotechnology.

Chapter 3

Fads and Hype in Technology: The Sargasso Sea of "Some Day Soon"

Peter Coffee

Is nanotechnology a fad? Prospective investors, or prospective research sponsors, want to know—and it's not too early in the life cycle of the field to ask, and answer, this question. First, though, let's be a little more specific about the ways that nanotechnology might disappoint us.

- It could be another cold fusion, with its germ of potential utility vastly outweighed by its hype—and with fractious scientific infighting scaring capital away to pursue more plausible pathways.
- It could be another synthetic-fuels or bubble-memory scenario, backed by valid science but not economically competitive with established approaches—or with their surprisingly long-lived processes of continuing improvement.
- It could be another artificial intelligence, a label that's reasonably defined as what we haven't figured out how to do—because once we can do it, it's assimilated into computer science, operations research, game theory, or some other respectable discipline.

It's reasonable to wonder whether the concepts that sail together under the "nanotechnology" label are heading for a new world of efficient use of energy and mass—or whether they're destined to circle indefinitely around a Sargasso Sea of "some day soon."

KNOWING SUCCESS WHEN WE SEE IT

In this chapter we look at some fundamental reasons nanotechnology represents a virtuous circle, in which every development enables additional progress, rather than sending us down a long and costly road with no real destination.

But just as *fad* is too general a word for how things might go wrong, *nanotechnology* is also so big an umbrella that it's bound to harbor both realities and fantasies beneath it. The signal-to-noise ratio—the proportion of reliable information to bad—may always be less than we'd like, because the upper bound on what nanotechnology could someday mean is a compelling vision.

Moreover, as we learn to do specific things, and as that knowledge makes the move from concept to practice, it will be natural for people to devise more specific labels for what they're doing successfully. When people are trying to attract resources to present-day efforts, they want to distance themselves from speculative visions of what nanotechnology might someday enable. Paradoxically, this need of researchers to define their own niches could make it seem as if the term *nanotechnology* has vanished from practical discussions.

The more ways that nanotechnology succeeds, the sooner this could happen.

Disappearance of the word would not mean, though, that nanotechnology was a fad. We can't equate the success of a technology with the ubiquity or the longevity of its label. Some technology labels come and go, whereas others survive for decades: We do well to recognize that a label sometimes disappears through success rather than failure.

The labels that hang around forever may describe what we merely wish we could invent—for example, "perpetual motion." Other labels fade from use, though, precisely because the thing they describe has become the mainstream. We can already see, for example, the beginnings of the disappearance of the adjectives in *cellular phone, digital camera,* and perhaps, in a few more years, *hybrid car.*

To today's teenagers, the first item is just a phone; to their younger brothers and sisters, the second is just a camera. And by the time a child born tomorrow has learned to drive, the third will just be a car. Will it seem quaint, in 2020 or perhaps as soon as 2010, to say that a product incorporates nanotechnology components or that it's built using nanotechnology methods? Will this term seem as dated as a label on a 1960s or 1970s radio that proudly proclaims it to be "solid state," trumpeting its use of transistors rather than tubes?

Yes, in all likelihood it will.

Nanotechnology is not a fad, because it is a trend: It's being adopted, not defensively by firms afraid of being left behind, but aggressively by companies that see what it can do—and by communities of research and application that are meeting long-felt needs through nanoscale control of ingredients and processes.

Even if nanotechnology does have its roots in the singular vision of a Richard Feynman or a K. Eric Drexler, that tree quickly leafs out into branches that extend into every industry. One recalls the conversation from the 1967 movie *The Graduate,* in which an adult counsels a newly minted college graduate that the one word that matters is *plastics:* Those who saw the new things that plastics enabled, and did them, have most likely outperformed those who merely focused on the materials themselves. So it will be with nanotechnology: Those who apply it will achieve greater returns than those who merely enable and deliver it.

The nanotechnology components of finished products such as flat-panel displays will represent a fraction of the product's value; far more significant will be the radical reduction of the up-front capital cost of manufacturing such devices, lowering barriers to entry and thus changing the entire landscape of that marketplace.

THE NATURE OF HYPE

The flip side of a fad, though, is the problem of hype: excessive chatter about a technology that generates unreasonable expectations. It would be one thing if hype were merely a disconnected phenomenon, like an instrument that gives wildly inaccurate readings without any possibility of actually changing what it purports to measure. In the realms of research and finance, though, hype can actually drive the best teams away from a field.

Why? A pragmatic researcher may choose to avoid the no-win proposition of either failing, on the one hand, or succeeding only to find that people are unimpressed with what they had been falsely told was sure to happen sooner or later. Computer recognition of unconstrained human speech is a good example: Even the best actual products are simply no match for what people have been seeing on science-fiction TV shows for decades.

Similarly, a finance team may want to avoid being tarred with the brush of only going after the obvious opportunities, even when it takes real talent to see the genuine prospects through the smoke of popular misperceptions.

Researchers from Gartner, an analysis and research company specializing in the information technology industry, have even defined a "hype cycle," which begins with a "technology trigger" that creates significant interest. One might pick any of several such triggers for the nanotech manifestation of this cycle, but K. Eric Drexler's 1986 novel *Engines of Creation* is probably a document that can be cast in this role.

Today's environment of commercial technology reporting and forecasting, venture capital (VC) looking for opportunity, and the conference and lecture circuit creates a volatile mix: A small injection of the fuel of technical potential is finding a vast oxidizing atmosphere of those who have an interest in anointing The Next Big Thing. The result is a hot but short-lived explosion of inflated expectations.

Investors may almost throw money at start-ups whose success in attracting capital may have little to do with their founders' actual skills as developers and implementers. People in the venture capital industry, observes David Eastman, General Partner at Prospector Equity Capital, have "the bad habit of sometimes jumping into and out of the latest fad together. You've seen it happen with disk drives, optical networking and network storage." Eastman has warned, "VCs will fund three or four times as many players as can possibly survive in a hot sector; when the hockey stick projections don't develop in the expected time frame, many will abandon the sector, often just as it's starting to develop into a real business."

The subsequent disillusionment may last much longer than the initial flush of interest, as those who expected too much seek to wind up with at least an aura of hard-earned wisdom by narrating their lessons learned. This is when good technologies may find themselves starved of needed resources for legitimate, moderately paced refinement, and this is what nanotechnology proponents might justifiably have feared only a few years ago.

What keeps a technology out of that pit, though, is the combined effects of many different players finding many different applications of the kinds described here and in other chapters. The need is for applications in which fundamental barriers of physical possibility, or thickets of diminishing returns on increasing investments, make it clear that something better is needed. The present ferment of nanoscale technologies is being driven by such applications, and therefore by actual science rather than mere marketing. Capital and talent are being pulled into play by the opportunity that exists to develop products that have solid advantages to offer. This is quite different from the dot-com boom, in which success was measured by hits on unprofitable Web sites.

The resulting "slope of enlightenment," as Gartner puts it, is gradual but important: It leads to an even less dramatic, but ultimately more valuable, "plateau of productivity" on which most of the money is made—even while the chattering techorati lament the imminent end of the era. Silicon-based semiconductors come to mind: Their displacement by gallium arsenide has been confidently predicted as both necessary and imminent for at least a decade, but it seems very little closer now than it did in the early 1990s.

Given the fundamental nature of atoms, perhaps the end of the nanotechnology era will not be chatted up so prematurely—since it's hard to imagine what will succeed it.

CLOUDING THE PICTURE

It can be hard, though, to see a clear picture of nanotechnology's bright future through the haze of arguments over terminology. Arguably, for example, current semiconductor manufacturing techniques operate at the nanoscale in that they meet the simple test of working with dimensions on the order of nanometers.

In the same way that a circuit designer needs an oscilloscope that has at least a 10-GHz bandwidth to accurately characterize a 1-GHz signal, a semiconductor fabrication line needs nanometer (or even subnanometer) accuracy to fabricate devices with feature sizes on the order of 20 nanometers thick, or about one thousandth the width of a human hair. That is the announced goal of Intel, for example, which hopes to produce flash memory chips on that scale around 2012.[1]

Others, though, insist that "real" nanotechnology is the building of devices that give users molecular-level control of the content and arrangement of the components of the device, as opposed to merely working with bulk materials at nanometer scale. Consider the building of carbon nanotubes, which are cylinders made up of pentagonal rings about a nanometer in diameter. Building carbon nanotubes is a molecule-by-molecule process—one that has yet to achieve the simplicity and consistency, and therefore the economy, of growing a monolithic semiconductor-grade silicon crystal.

It's certain that key nanotechnology goals depend on achieving this level of control, but progress is being made along these lines. Work published in November 2004, for example, has described the use of artificially constructed DNA molecules to orient and place carbon nanotubes between electrode pairs to create transistors.[2]

This kind of synergy—that is, the use of one kind of molecular engineering to carry out another kind of molecular construction process—seems likely to be a common theme of successful nanotechnology developments.

Moreover, mass-market manufacturers such as South Korea's Samsung are already applying the properties of nanoscale materials such as carbon nanotubes to their next-generation products. Already working in prototype form, for example, are flat-panel TV screens, known as "field emission displays," that are on track for mass-market availability by the end of 2006.[3] Functioning as an array of precise and compact electron guns bombarding a phosphor screen, a nanotube backplane allows a display designer to combine superior brightness, sharpness, and power consumption characteristics with the convenient flat-panel form.

Any consumer electronics maker that does not match Samsung's aggressive investment in this technology is rehearsing for a replay of the fate that befell competitors of Sony and Panasonic in the 1960s, when those companies essentially created the pocket-sized transistor radio as the icon of the solid-state revolution.[4] Note well that although Sony was not the first to offer an all-transistor radio, its 1955 product was the first to come from a company that made its own transistors. Subsequent innovations, such as Sony's use of vertical-field-effect devices in FM-stereo receivers came from the same focus on core technology. Samsung's Advanced Institute of Technology (south of Seoul) represents a similar commitment of energy and resources.

It's also vital to recognize that nanomaterial applications stem from the many different characteristics of these unfamiliar substances. The electrical properties of carbon nanotubes are one thing; their thermal properties—specifically, the high thermal conductivity that could make them a useful addition to the heat-transferring "thermal grease" that's applied between computer microprocessors and their heat sinks—are another.[5]

Difficulties in fabrication consistency that might cripple one application—for example, one that depends on a consistent nanotube length—could be irrelevant in a task, such as heat conduction, that depends only on aggregate properties. Other such aggregate-property impacts of nanomaterials include improved air retention in tennis balls, better stain resistance in fabrics, longer shelf life and better skin-penetration characteristics for cosmetic and physical-therapy products, and reduced degradation of meat and cheese products prior to sale.[6] These are not merely potential applications, but are actual uses of mass-produced nanomaterials that are now being incorporated into consumer products. For example, nearly half of Dockers clothing uses nanotechnology-based fabric treatments that resist stains or improve perspiration handling.

It's also important to note that nanomaterials are interesting, not merely for their size and their average properties, but for their potential of achieving unprecedented levels of consistency in many physical characteristics that were previously far below the "noise floor" of conventional manufacture. In the same way that lasers enabled new processes by offering a coherent source of perfectly monochromatic light, a nanotechnological process can deliver what we might call coherent objects—for example, the so-called nanoshell spheres that can be precisely tailored to convert a particular wavelength of light into heat. When coated with a material that preferentially bonds to cancerous tumor cells, this property of nanoshells paves the way to a highly specific tumor-killing technique (described in *Cancer Letters* in June 2004) that has far fewer detrimental side effects than other cancer treatments.[7]

Even for those who insist on infinitesimally small mechanisms, built to order from molecular-scale building blocks, the nanotechnology practitioners are ready to deliver. Prototypes of 35-nanometer mechanical valves, potentially suited to applications such as ink delivery in ink-jet printers, were constructed in 2004 from assemblies of roughly 75,000 atoms: Components derived from this work could be seen in printers by 2008, and perhaps in drug-delivery systems before 2015, according to researchers at California Institute of Technology, who demonstrated the use of a silicon beam to pinch closed a carbon nanotube.[8]

A VIRTUOUS CIRCLE

Technologies can succeed only when they combine appropriate control of energy, materials, design, measurement, fabrication, and marketing. Absent any of these, one has at best a laboratory demonstration; at worst, a costly or even fraudulent disappointment.

Nanotechnology, though, has all these elements at the ready. What's more, nanotechnology represents a practical solution to fundamental limits being encountered in all these areas by existing approaches.

Nanotechnology therefore has vastly more friends than enemies: There is really no industry that's likely to perceive nanotechnology as a threat. Rather, it represents the prospect of offering many industries their next generation of improved products that deliver more capability—or novel, previously unobtainable capability—at a quite attractive price.

A virtuous circle can be kinked or broken, though, if those who offer a supposed technology improvement fail to understand the environment in which it will be used. On a much larger scale than the nanometer, it might look as if buttons (which can be lost) or zippers or drawstrings (which can break or jam) might be better replaced by Velcro fasteners on soldiers' battle uniforms, but a soldier's first reaction to a Velcro-flapped pocket might well be, "This makes way too much noise!"

Before manufacturers develop nanotechnology products based on what the developers think are superior features or properties, it's essential to study the ways that existing products are actually used—and the features or properties of those products that users never think to mention when asked what they like or dislike, because it's never occurred to them that things could be otherwise.

At the same time, those who bring new products to market must anticipate and address prospective buyers' objections—even when those objections arise from a buyer's combining past experience with inadequate understanding of future benefits. When vacuum tubes were being replaced by transistors in field communications equipment, one experienced technician complained to an engineer that a failed transistor needed a piece of test gear to identify it, unlike a blown tube, which could easily be recognized on sight. That's perfectly accurate, but it failed to appreciate the enormous reduction in the actual number of device failures that would be encountered in the field with solid-state devices.

Bringing nanoscale processes and products to market may depend on overcoming similarly incomplete understanding of their impact.

Technologies fail to gain traction in the marketplace when they require too great a discontinuous leap from existing modes of design and production. Nanotechnology has already succeeded, though, in building on existing manufacturing foundations; one example is the May 2004 production of carbon-nanotube computer memory cells on a high-volume semiconductor production line by LSI Logic, working with nanotechnology semiconductor pioneer Nantero Inc. in Woburn, Massachusetts.[9]

Perhaps equally important is that nanotechnology semiconductors don't depend on one single approach, such as Nantero's. Another company, Zettacore Inc., of Englewood, Colorado, is developing memory chips that use a chlorophyll-derived molecule having the ability to act as an electron-storage site. Still another approach has been demonstrated at Boston University, where a microscopic mechanical beam—in late-2004 tests, 8,000 nanometers long—can be made to flex in either of two directions, thereby storing a binary 1 or 0, the form in which computer data is stored and manipulated. When scaled down to 1,000-nanometer length, the resulting computer memory device is expected to store 100 gigabytes per square inch while operating at frequencies exceeding 1 GigaHertz, surpassing semiconductor memory in both respects.

These three approaches—one based on an engineered molecule, another on a biologically derived molecule, and yet another on a nanoscale implementation of a familiar physical behavior—demonstrate how broad the nanotechnology field really is.

Putting aside the *nano-* prefix, let's look for a moment at the many aspects that make up a technology. The technology of firearms is a well-studied example. Its foundations include the packaging and control of energy on the proper scale and in a convenient and reliable form. Without gunpowder that's stable enough to store but prompt and vigorous in detonating when desired, nothing else could happen.

Nanotechnology has that control of energy at its disposal—whether in the form of electrons, photons, or mechanical storage methods.

Firearms also depend on materials science: Simple materials could be used to construct bulky, failure-prone cannons, but a practical portable firearm demanded more sophisticated knowledge of materials behavior. Applications of nanomaterials that are already in the marketplace display this knowledge.

Finally, and most often noted when firearms are used as an example, there is the need for precise and reproducible methods of manufacture. This implies tools and techniques of measurement and fabrication. Nanoscale manufacture, in the purists' sense of controlling the arrangements of individual molecules or even atoms, implies at the very least the ability to resolve images of structures on this scale. It's one thing to measure bulk properties of materials and to estimate the degree of molecular order and purity that's been attained; it's another to observe, let alone manipulate, material fragments at this level of precision.

Progress in this area has been notable, though, and promising. FEI Co., in Hillsboro, Oregon, announced in 2004 its achievement of image resolution below the threshold of one Angstrom unit (0.1 nanometers)—that is, about the size of a single hydrogen atom.[10] The U.S. Department of Energy began work in 2000 on a microscope with half-Angstrom resolution, using an aberration-correcting assembly of multiple magnetic-field "lens" elements: As of the end of 2004, the first such unit at Lawrence Berkeley National Laboratory, in Berkeley, California, was expected to be operational by 2008.

WHEN SCIENCE GETS DOWN TO BUSINESS

If adolescence must come between infancy and maturity, then nanotechnology's maturity is perhaps on the way—because adolescent behavior is already starting to emerge. Scamsters and rip-off artists may already be entering the nanomaterials marketplace: Anecdotal reports have described, for example, the shipment of ordinary carbon soot to buyers who thought they were getting nanotubes, or the receipt of high-priced nanotube shipments containing tube-growth catalyst making up as much as one-third of the volume.

There are indications that the materials most prone to customer complaint are those, such as carbon nanotubes and fullerenes, in which precise molecular structure is everything. Other items, such as nanoporous materials, whose value is better approximated by statistically measurable behaviors, appear to be less of a problem. This suggests that precise measurements, standardized and reliable tests, and mutually agreed-upon standards for characterizing and verifying nanomaterial quality will be needed to make nanomaterials ready for full-time employment.

Another adolescent aspect of nanotechnology is the question of product safety. Nanoscale particles may be able to bypass many of the human body's long-standing defenses, such as the skin and the circulatory vessels' walls; toxicity issues may therefore arise. Car exhaust, toner particles from xerographic copiers, and drug molecules are already produced and used in many environments without dramatic ill effects being observed, but nanotechnology adopters must exercise responsibility and manage public relations to address both legitimate and exaggerated concerns.[11]

FAR FROM A FAD

We asked at the beginning of this chapter whether nanotechnology might be a kernel of utility buried in a heap of hype. It's clear that this is not the case. The problems that nanotechnology is already solving are fundamental; their solution is clearly valuable.

We asked whether the continued evolution of other approaches might make nanotechnology always interesting, but never attractive in the mass market. That scenario has already been left behind. Nanomaterials are not merely competitive, but are poised on the threshold of becoming the dominant choice for many tasks.

We asked whether nanotechnology would remain forever a theoretical or philosophical possibility, but always blocked from widespread use in practical manufacturing or other mainstream situations. Again, that question has already been answered by existing uses, such as fabric treatments and drug-delivery tools, with additional answers in the successful prototyping of new types of nanoscale mechanical component or full-scale consumer electronics devices.

Not a fad, but a trend; not a field in itself, but a transformation in nearly every field of technology. If there's bad news, it's that nanotechnology may be the next commodity: The mere ability to fabricate and produce basic materials at the nanoscale may command unexciting profit margins beyond the time frame of the 2010s, or at latest the 2020s.

The good news, though, is that the resulting new technology foundations will enable explosions of new value on the upper levels of industry, medicine, and lifestyle.

REFERENCES

1. http://www.reed-electronics.com/eb-mag/article/CA475441?industryid=2116.
2. http://www.technologyreview.com/articles/04/12/rnb_120304.asp?p=1.
3. http://www.technologyreview.com//articles/04/11/mann1104.asp.
4. http://eetimes.com/special/special_issues/millennium/companies/sony.html.
5. http://news.com.com/Can+nanotubes+keep+PCs+cool/2100-7337_3-5166111.html?tag=nl.
6. http://www.usatoday.com/tech/news/nano/2004-11-08-nano-on-the-move_x.htm.
7. http://sciencentral.com/articles/view.php3?article_id=218392390.
8. http://www.trnmag.com/Stories/2004/102004/Mechanical_valve_design_goes_nano_Brief_102004.html.
9. http://www.trnmag.com/Stories/2004/111704/Nanomechanical_memory_demoed_111704.html.
10. http://news.com.com/Superpowered+micro-scope+could+help+focus+nanotech+dream/2100-7337_3-5471939.html.
11. http://www.dailycal.org/article.php?id=17119.

SECTION TWO

The Players

Chapter 4

Nanotechnology Commercialization: Transcending Moore's Law with Molecular Electronics and Nanotechnology

Steve Jurvetson

The history of technology is one of disruption and exponential growth, epitomized in Moore's Law and generalized to many basic technological capabilities that are compounding independently from the economy. More than a niche subject of interest only to computer chip designers, the continued march of Moore's Law will affect all sciences, just as nanotech will affect all industries. (We define Moore's Law in some detail a bit later in this chapter .)

Thinking about Moore's Law in the abstract provides a framework for predicting the future of computation and the transition to a new substrate: molecular electronics. An analysis of progress in molecular electronics provides a detailed example of the commercialization challenges and opportunities common to many nanotechnologies.

TECHNOLOGY EXPONENTIALS

Despite a natural human tendency to presume linearity, accelerating change from positive feedback is a common pattern in technology and evolution. We are now crossing a threshold where the pace of disruptive shifts is no longer intergenerational; instead, it begins to have a meaningful impact over the span of careers and eventually over product cycles.

As early-stage venture capitalists (VCs), we look for disruptive businesses run by entrepreneurs who want to change the world. To be successful, we must identify technology waves early and act upon those beliefs. At Draper Fisher Jurvetson (DFJ), we believe that nanotech is the next great technology wave, the nexus of scientific innovation that revolutionizes most industries and indirectly affects the fabric of society. Historians will look back on the upcoming epoch as having no less portent than the Industrial Revolution.

The aforementioned are some long-term trends. Today, from a seed-stage VC perspective (with a broad sampling of the entrepreneurial pool), we are seeing more innovation than ever before. And we are investing in more new companies than ever before.

In the medium term, disruptive technological progress is relatively decoupled from economic cycles. For example, for the past 40 years in the semiconductor industry, Moore's Law has not wavered in the face of dramatic economic cycles. Ray Kurzweil's abstraction of Moore's Law (from its focus on transistors to a broader focus on computational capability and storage capacity) shows an uninterrupted exponential curve for more than 100 years, again without perturbation during the Great Depression or the world wars. Similar exponentials can be seen in Internet connectivity, medical imaging resolution, genes mapped, and solved 3-D protein structures. In each case, the level of analysis is not products or companies, but basic technological capabilities.

Kurzweil has summarized the exponentiation of our technological capabilities, and our evolution, with this near-term shorthand: The next 20 years of technological progress will be equivalent to that of the entire twentieth century. For most of us, who do not recall what life was like 100 years ago, the metaphor is a bit abstract. In the early 1900s, in the United States, there were only 144 miles of paved road, and most Americans (more than 94%) were born at home, without a telephone, and never graduated from high school. Most (more than 86%) did not have a bathtub at home nor reliable access to electricity. Consider how much technology-driven change has compounded over the past century, and consider that an equivalent amount of progress will occur in one human generation, by 2020. It boggles the mind, until one dwells on genetics, nanotechnology, and their intersection. Exponential progress perpetually pierces the linear presumptions of our intuition. "Future shock" is no longer on an intergenerational time scale.

The history of humanity is that we use our knowledge and technical expertise to build better tools and expand the bounds of our learning. We are entering an era of exponential growth in our capabilities in biotech, molecular engineering, and computing. The cross-fertilization of these formerly discrete domains compounds our rate of learning and our engineering capabilities across the spectrum. With the digitization of biology and matter, technologists from myriad backgrounds can decode and engage the information systems of biology as never before. This inspires new approaches to bottom-up manufacturing, self-assembly, and the development of layered, complex systems.

MOORE'S LAW

Moore's Law is commonly reported as a doubling of transistor density every 18 months. But this is not something that Gordon Moore, co-founder of Intel, has ever said. It is a nice blending of his two predictions; in 1965, he predicted an annual doubling of transistor counts in the most cost-effective chip, and he revised it in 1975 to every 24 months. With a little hand waving, most reports attribute 18 months to Moore's Law, but there is quite a bit of variability.

The popular perception of Moore's Law is that computer chips are compounding in their complexity at a nearly constant per-unit cost. This is one of the many abstractions of Moore's Law, and it relates to the compounding of transistor density in two dimensions. Others relate to speed (the signals have less distance to travel) and to computational power (speed × density).

So as not to miss the long-term trend while sorting out the details, we will focus on the 100-year abstraction of Moore's Law. But we should digress for a moment to underscore the importance of continued progress in Moore's Law to a broad set of industries.

The Importance of Moore's Law

Moore's Law drives chips, communications, and computers and has become the primary driver in drug discovery and bioinformatics, medical imaging and diagnostics. Over time, the lab sciences become information sciences, modeled on a computer rather than trial-and-error experimentation.

The NASA Ames Research Center shut down its wind tunnels this year. Because Moore's Law provided enough computational power to model turbulence and airflow, there was no longer a need to test iterative physical design variations of aircraft in the wind tunnels, and the pace of innovative design exploration dramatically accelerated.

Pharmaceutical giant Eli Lilly processed 100 times fewer molecules this year than it did 15 years ago, but its annual productivity in drug discovery did not drop proportionately; it went up over the same period. "Fewer atoms and more bits" is Lilly's coda.

Accurate simulation demands computational power, and once a sufficient threshold has been crossed, simulation acts as an innovation accelerant compared with physical experimentation. Many more questions can be answered per day.

Recent accuracy thresholds have been crossed in diverse areas, such as modeling the weather (predicting a thunderstorm six hours in advance) and automobile collisions (a relief for the crash test dummies), and the thresholds have yet to be crossed for many areas, such as protein folding dynamics.

Unless you work for a chip company and focus on fab-yield optimization, you do not care about transistor counts. Integrated circuit customers do not buy transistors. Consumers of technology purchase computational speed and data storage density. When recast in these terms, Moore's Law is no longer a transistor-centered metric, and this abstraction allows for longer-term analysis.

The exponential curve of Moore's Law extends smoothly back in time for more than 100 years, long before the invention of the semiconductor. Through five paradigm shifts—such as electromechanical calculators and vacuum tube computers—the computational power that $1,000 buys has doubled every two years. For the past 30 years, it has been doubling every year.

Each horizontal line on the logarithmic graph given in Figure 4–1 represents a 100 times improvement. A straight diagonal line would be an exponential, or geometrically compounding, curve of progress. Kurzweil plots a slightly upward curving line—a double exponential.

Each dot represents a human drama, although the humans did not realize that they were on a predictive curve. Each dot represents an attempt to build the best computer using the tools of the day. Of course, we use these

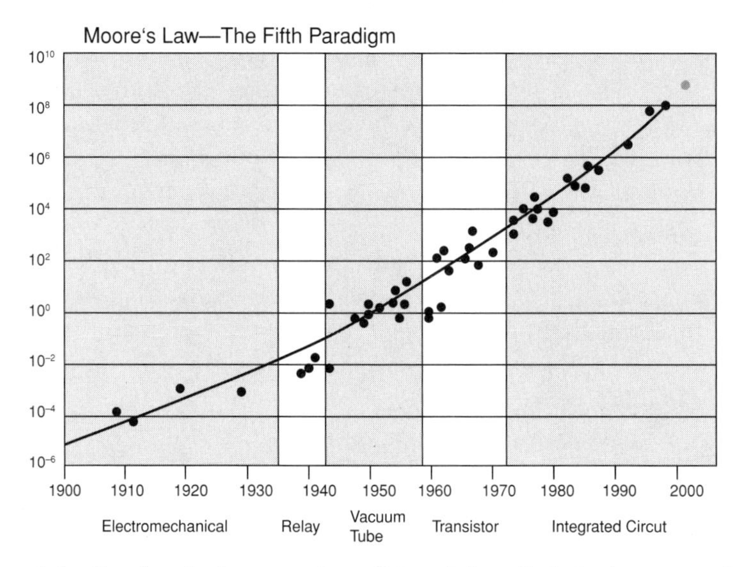

Figure 4–1 One-hundred-year version of Moore's Law. Each dot is a computing machine. (Source: Ray Kurzweil.)

computers to make better design software and algorithms (formulas) for controlling manufacturing processes. And so the progress continues.

One machine was used in the 1890 census; another one cracked the Nazi Enigma cipher in World War II; still another one predicted Eisenhower's win in the presidential election. And there is the Apple II, and the Cray 1, and just to make sure the curve had not petered out recently, I looked up the cheapest PC available for sale on Wal-Mart.com, and that is the gray dot that I have added to the upper-right corner of the graph.

And notice the relative immunity to economic cycles. The Great Depression and the world wars and various recessions do not introduce a meaningful delay in the progress of Moore's Law. Certainly, the adoption rates, revenue, profits, and inventory levels of the computer companies behind the various dots on the graph may go through wild oscillations, but the long-term trend emerges nevertheless.

Any one technology, such as the complementary metal oxide semiconductor (CMOS) transistor, follows an elongated S-curve of slow progress during initial development, upward progress during a rapid adoption phase, and then slower growth from market saturation over time. But a more generalized capability, such as computation, storage, or bandwidth, tends to follow a pure exponential—bridging a variety of technologies and their cascade of S-curves.

If history is any guide, Moore's Law will continue and will jump to a different substrate than CMOS silicon. It has done so five times in the past and will need to again in the future.

Problems with the Current Paradigm

Gordon Moore has chuckled at those in past decades who predicted the imminent demise of Moore's Law. But the traditional semiconductor chip is finally approaching some fundamental physical limits. Moore recently admitted that Moore's Law, in its current form, with CMOS silicon, will reach its limit in 2017.

One of the problems is that the chips are operating at very high temperatures. This provides the impetus for chip-cooling companies to provide a breakthrough solution for removing 100 watts per square centimeter. In the long term, the paradigm must change.

Another physical limit is the atomic limit—the indivisibility of atoms. Intel's current gate oxide is 1.2nm thick. Intel's 45nm process is expected to have a gate oxide that is only three atoms thick. It is hard to imagine many more doublings from there, even with further innovation in insulating materials. Intel has recently announced a breakthrough in a nanostructured gate oxide (high-k dielectric) and metal contact materials that should enable the 45nm node to come online in 2007. None of the industry participants has a CMOS road map for the next 50 years.

A major issue with thin gate oxides, and one that will also come to the fore with high-k dielectrics, is quantum mechanical tunneling. As the oxide becomes thinner, the gate current can approach and even exceed the channel current to the point that the transistor cannot be controlled by the gate.

Another problem is the escalating cost of a semiconductor fabrication plant (called a *fab*), a cost that is doubling every three years, a phenomenon dubbed Moore's Second Law. Human ingenuity keeps shrinking the CMOS transistor, but with increasingly expensive manufacturing facilities—currently $3 billion per fab.

A large component of fab cost is the lithography equipment that is used to pattern the wafers with successive submicron layers. Nanoimprint lithography can dramatically lower cost and leave room for further improvement from the field of molecular electronics.

We have been investing in a variety of companies that are working on the next paradigm shift to extend Moore's Law beyond 2017. One near-term extension to Moore's Law focuses on the cost side of the equation. Imagine rolls of wallpaper embedded with inexpensive transistors. One of our companies deposits traditional transistors at room temperature on plastic, a much cheaper bulk process than growing and cutting crystalline silicon ingots.

MOLECULAR ELECTRONICS

The primary contender for the post-silicon computation paradigm is molecular electronics, a nanoscale alternative to the CMOS transistor. Eventually, molecular switches will revolutionize computation by scaling into the third dimension—overcoming the planar deposition limitations of CMOS. Initially, these switches will substitute for the transistor bottleneck that results from a standard silicon process using standard external input/output interfaces.

For example, Nantero, a nanotech firm based in Woburn, Massachusetts, employs carbon nanotubes suspended above metal electrodes on silicon to create high-density nonvolatile memory chips (the weak Van der Waals bond can hold a deflected tube in place indefinitely with no power drain). Carbon nanotubes are small (approximately 10 atoms wide), 30 times as strong as steel at one-sixth the weight, and they perform the functions of wires, capacitors, and transistors with better speed, power, density, and cost. Cheap nonvolatile memory enables important advances, such as "instant-on" PCs.

Other companies, such as Hewlett-Packard and ZettaCore, are combining organic chemistry with a silicon substrate to create memory elements that self-assemble using chemical bonds that form along prepatterned regions of exposed silicon.

There are several reasons molecular electronics is the next paradigm for Moore's Law:

- **Size:** Molecular electronics has the potential to dramatically extend the miniaturization that has driven the density and speed advantages of the integrated circuit (IC) phase of Moore's Law. In 2002, using a scanning tunneling microscope (STM) to manipulate individual carbon monoxide molecules, IBM built a three-input sorter by arranging those molecules precisely on a copper surface. It is 260,000 times as small as the equivalent circuit built in the most modern chip plant. For a memorable sense of the difference in scale, consider a single drop of water. There are more molecules in a single drop of water than in all the transistors ever built. Think of the transistors in every memory chip and every processor ever built; there are about 100 times as many molecules in a drop of water. Certainly, water molecules are small, but an important part of the comparison depends on the 3-D volume of a drop. Every IC, in contrast, is a thin veneer of computation on a thick and inert substrate.
- **Power:** One of the reasons that transistors are not stacked into 3-D volumes today is that the silicon would melt. The inefficiency of the modern transistor is staggering. It is much less efficient at its task than the internal combustion engine. The brain provides an existing proof of what is possible; it is 100 million times as efficient in power and calculation as our best processors. Sure, it is slow (less than 1 kHz), but it is massively interconnected (with 100 trillion synapses between 60 billion neurons), and it is folded into a 3-D volume. Power per calculation will dominate clock speed as the metric of merit for the future of computation.
- **Manufacturing cost:** Many of the molecular electronics designs use simple spin coating or molecular self-assembly of organic compounds. The process complexity is embodied in the synthesized molecular structures, and so they can literally be splashed on to a prepared silicon wafer. The complexity is not in the deposition nor the manufacturing process nor the systems engineering. Much of the conceptual difference of nanotech products derives from a biological metaphor: Complexity builds from the bottom up and pivots about conformational changes, weak bonds, and surfaces. It is not engineered from the top down with precise manipulation and static placement.

- **Low-temperature manufacturing:** Biology does not tend to assemble complexity at 1,000 degrees in a high vacuum. It tends to work at room temperature or body temperature. In a manufacturing domain, this opens the possibility of using cheap plastic substrates instead of expensive silicon ingots.
- **Elegance:** In addition to these advantages, some of the molecular electronics approaches offer elegant solutions to nonvolatile and inherently digital storage. We go through unnatural acts with CMOS silicon to get an inherently analog and leaky medium to approximate a digital and nonvolatile abstraction that we depend on for our design methodology. Many of the molecular electronic approaches are inherently digital, and some are inherently nonvolatile.

Other research projects, from quantum computing to using DNA as a structural material for directed assembly of carbon nanotubes, have one thing in common: They are all nanotechnology.

THE COMMERCIALIZATION OF NANOTECHNOLOGY

Nanotech is often defined as the manipulation and control of matter at the nanometer scale (critical dimensions of 1 to 100nm). It is a bit unusual to describe a technology by a length scale. We certainly didn't get very excited by "inch-o technology." As venture capitalists, we start to get interested when there are unique properties of matter that emerge at the nanoscale and that cannot be exploited at the macroscale world of today's engineered products. We like to ask the start-ups that we are investing in, "Why now? Why couldn't you have started this business ten years ago?" The responses of our nanotech start-ups have a common thread: Recent developments in the capacity to understand and engineer nanoscale materials have enabled new products that could not have been developed at larger scale.

Various unique properties of matter are expressed at the nanoscale and are quite foreign to our "bulk statistical" senses (we do not see single photons or quanta of electric charge; we feel bulk phenomena, like friction, at the statistical or emergent macroscale). At the nanoscale, the bulk approximations of Newtonian physics are revealed for their inaccuracy and give way to quantum physics. Nanotechnology is more than a linear improvement with scale; everything changes. Quantum entanglement, tunneling, ballistic transport, frictionless rotation of superfluids, and several other phenomena have been regarded as "spooky" by many of the smartest scientists, even Einstein, upon first exposure.

For a simple example of nanotech's discontinuous divergence from the "bulk" sciences, consider the simple aluminum soda can. If you take the inert aluminum metal in that can and grind it down into a powder of 20–30nm particles, it will spontaneously explode in air. It becomes a rocket fuel catalyst. In other words, the energetic properties of matter change at that scale. The surface-area-to-volume ratios become relevant, and even the distances between the atoms in a metal lattice change from surface effects.

Innovation from the Edge

Disruptive innovation, the driver of growth and renewal, occurs at the edge. In start-ups, innovation occurs out of the mainstream, away from the warmth of the herd. In biological evolution, innovative mutations take hold at the physical edge of the population, at the edge of survival. In complexity theory, structure and complexity emerge at the edge of chaos—the dividing line between predictable regularity and chaotic indeterminacy. And in science, meaningful disruptive innovation occurs at the interdisciplinary interstices between formal academic disciplines.

Herein lies much of the excitement about nanotechnology: in the richness of human communication about science. Nanotech exposes the core areas of overlap in the fundamental sciences, the place where quantum physics and quantum chemistry can cross-pollinate with ideas from the life sciences.

Over time, each of the academic disciplines develops its own proprietary systems vernacular that isolates it from neighboring disciplines. Nanoscale science requires scientists to cut across the scientific languages to unite the isolated islands of innovation. As illustrated in Figure 4–2, nanotech is the nexus of the sciences.

In academic centers and government laboratories, nanotech is fostering new discussions. At Stanford, UCLA, Duke, and many other schools, the new nanotech buildings are physically located at the symbolic hub of the schools of engineering, computer science, and medicine.

Nanotech is the nexus of the sciences, but outside the sciences and research itself, the nanotech umbrella conveys no business synergy whatsoever. The marketing, distribution, and sales of a nanotech solar cell, memory chip, or drug delivery capsule will be completely different from each other and will present few opportunities for common learning or synergy.

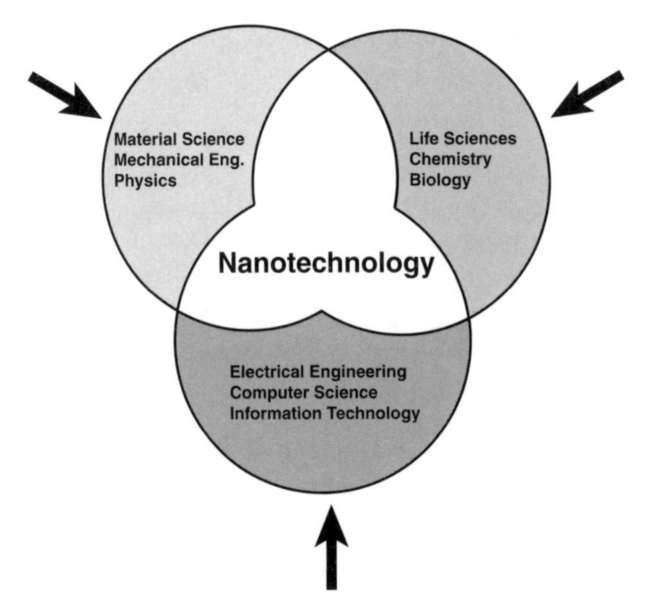

Figure 4–2 Nanotech is the nexus of the sciences.

Market Timing

As an umbrella term for a myriad of technologies spanning multiple industries, *nanotech* will eventually disrupt these industries over different time frames—but most are long-term opportunities. Electronics, energy, drug delivery, and materials are areas of active nanotech research today. Medicine and bulk manufacturing are future opportunities. The National Science Foundation predicts that nanotech will have a trillion-dollar impact on various industries within 15 years.

Of course, if one thinks far enough in the future, every industry eventually will be revolutionized by a fundamental capability for molecular manufacturing—from the inorganic structures to the organic and even the biological. Analog manufacturing will become digital, engendering a profound restructuring of the substrate of the physical world.

Futuristic predictions of potential nanotech products have a near-term benefit. They help attract some of the best and brightest scientists to work on hard problems that are stepping-stones to the future vision. Scientists relish exploring the frontier of the unknown, and nanotech embodies the tangible metaphor of the inner frontier.

Given that much of the abstract potential of nanotech is a question of "when" and not "if," the challenge for the venture capitalist is one of market timing. When should we be investing, and in which subsectors? It is as if we need to pull the sea of possibilities through an intellectual filter to tease apart the various segments into a time line of probable progression. That is an ongoing process of data collection (for example, the growing pool of business plan submissions), business and technology analysis, and intuition.

Two touchstone events for the scientific enthusiasm for the timing of nanotech were the decoding of the human genome and the dazzling visual images output by the scanning tunneling microscope (such as the arrangement of individual xenon atoms into the IBM logo). These events represent the digitization of biology and matter—symbolic milestones for accelerated learning and simulation-driven innovation.

More recently, nanotech publication has proliferated, as in the early days of the Internet. In addition to the popular press, the number of scientific publications on nanotech has grown by a factor of 10 in the past ten years. According to the U.S. Patent and Trademark Office (USPTO), the number of nanotech patents granted each year has skyrocketed by a factor of 3 in the past seven years. Ripe with symbolism, IBM has more lawyers working on nanotech than engineers.

With the recent codification of the National Nanotech Initiative into law, federal funding will continue to fill the pipeline of nanotech research. With $847 million earmarked for 2004, nanotech was a rarity in the tight budget process; it received more funding than was requested. Now nanotech is second only to the space race for federal funding of science. And the United States is not alone in funding nanotechnology. Unlike many previous technological areas, we aren't even in the lead; Japan outspends the United States each year on nanotech research. In 2003, the U.S. government spending was one-fourth of the world total.

Federal funding is the seed corn for nanotech entrepreneurship. All of our nanotech portfolio companies are spin-offs (with negotiated intellectual property [IP] transfers) from universities or government labs, and all got their start with federal funding. Often these companies need specialized equipment and expensive laboratories to do the early tinkering that will germinate a new breakthrough. These are typically lacking in the proverbial entrepreneur's garage.

Corporate investors have discovered a keen interest in nanotechnology, with internal R&D, external investments in start-ups, and acquisitions of promising companies, such as chipmaker AMD's recent acquisition of Coatue, a molecular electronics company.

Despite all this excitement, there are a fair number of investment dead ends, and so we continue to refine the filters we use in selecting companies to back. All entrepreneurs want to present their businesses as fitting an appropriate time line to commercialization. How can we guide our intuition to determine which of these entrepreneurs are right?

The Question of Vertical Integration

Nanotech involves the reengineering of the lowest-level physical layer of a system, and so a natural business question arises: How far forward do you need to vertically integrate before you can sell a product on the open market? For example, in molecular electronics, if you can ship a DRAM-compatible chip, you have found a horizontal layer of standardization, and further vertical integration is not necessary. If you have an incompatible 3-D memory block, you may have to vertically integrate to the storage subsystem level, or farther, to bring a product to market. That may require that you form industry partnerships, and it will, in general, take more time and money as change is introduced farther up the product stack. Three-dimensional logic with massive interconnectivity may require a new computer design and a new form of software; this would take the longest to commercialize. And most start-ups on this end of the spectrum would seek partnerships to bring their vision to market. The success and timeliness of that endeavor will depend on many factors, including IP protection, the magnitude of improvement, the vertical tier at which that value is recognized, the number of potential partners, and the needed degree of tooling and other industry accommodations.

Product development time lines are impacted by the cycle time of the R&D feedback loop. For example, outdoor lifetime testing for organic light-emitting diodes (LEDs) will take longer than in silicon simulation spins of digital products. If the product requires partners in the R&D loop or multiple nested tiers of testing, it will take longer to commercialize.

The Interface Problem

As we think about the start-up opportunities in nanotechnology, an uncertain financial environment underscores the importance of market timing and revenue opportunities over the next five years. Of the various paths to nanotech, which of them are 20-year quests in search of a government grant, and which are market-driven businesses that will attract venture capital? Are there co-factors of production that require a whole industry to be in place before a company ships products?

As a thought experiment, imagine that I could hand you today any nanotech marvel of your design—a molecular machine as advanced as you would like. What would it be? A supercomputer? A bloodstream submarine? A matter compiler capable of producing diamond rods or arbitrary physical objects? Pick something.

Now imagine some of the complexities: Did it blow off my hand as I offered it to you? Can it autonomously move to its intended destination? What is its energy source? How do you communicate with it?

These questions draw the interface problem into sharp focus: Does your design require an entire nanotech industry to support, power, and interface to your molecular machine? As an analogy, imagine that you have one of the latest Intel Pentium processors. How would you make use of the Pentium chip? You then need to wire-bond the chip to a larger lead frame in a package that connects to a larger printed circuit board, fed by a bulky power supply that connects to the electrical power grid. Each of these successive layers relies on its larger-scale precursors (which were developed in reverse chronological order), and the entire hierarchy is needed to access the potential of the microchip.

Where Is the Scaling Hierarchy for Molecular Nanotech?

To cross the interface chasm, today's business-driven paths to nanotech diverge into two strategies: the biologically inspired bottom-up path, and the top-down approach of the semiconductor industry. The developers of nonbiological microelectromechanical systems (MEMS) are addressing current markets in the micro world while pursuing an ever-shrinking spiral of miniaturization that builds the relevant infrastructure tiers along the way. Not surprisingly, this path is very similar to the one that has been followed in the semiconductor industry, and many of its adherents see nanotech as inevitable but in the distant future.

On the other hand, biological manipulation presents numerous opportunities to effect great change in the near term. Drug development, tissue engineering, and genetic engineering are all powerfully impacted by the molecular manipulation capabilities available to us today. And genetically modified microbes, whether by artificial evolution or directed gene splicing, give researchers the ability to build structures from the bottom up.

The Top-Down "Chip Path"

This path is consonant with the original vision of physicist Richard Feynman (in a 1959 lecture at Caltech) of the iterative miniaturization of our tools down to the nanoscale. Some companies are pursuing the gradual shrinking of semiconductor manufacturing technology from the MEMS of today into the nanometer domain of nanoelectromechanical systems (NEMS).

MEMS technologies have already revolutionized the automotive industry with air-bag sensors, and the printing sector with ink-jet nozzles, and they are on track to do the same in medical devices and photonic switches for communications and mobile phones. In-StatJMDR forecasts that the $4.7 billion in MEMS revenue in 2003 will grow to $8.3 billion by 2007. But progress is constrained by the pace (and cost) of the semiconductor equipment industry, and by the long turnaround time for fab runs.

Many of the nanotech advances in storage, semiconductors, and molecular electronics can be improved, or in some cases enabled, by tools that allow for the manipulation of matter at the nanoscale. Here are three examples:

- **Nanolithography:** Molecular Imprints is commercializing a unique imprint lithographic technology developed at the University of Texas at Austin. The technology uses photo-curable liquids and etched quartz plates to dramatically reduce the cost of nanoscale lithography. This lithography approach, recently added to the ITRS Roadmap, has special advantages for applications in the areas of nanodevices, MEMS, microfluidics, and optical components and devices, as well as molecular electronics.
- **Optical traps:** Arryx has developed a breakthrough in nanomaterial manipulation. Optical traps generate hundreds of independently controllable laser tweezers that can manipulate molecular objects in 3-D (move, rotate, cut, place), all from one laser source passing through an adaptive hologram. The applications span from cell sorting, to carbon nanotube placement, to continuous material handling. They can even manipulate the organelles inside an unruptured living cell (and weigh the DNA in the nucleus).
- **Metrology:** Imago's LEAP atom probe microscope is being used by the chip and disk drive industries to produce 3-D pictures that depict both the chemistry and the structure of items on an atom-by-atom basis. Unlike traditional microscopes, which zoom in to see an item on a microscopic level, Imago's nanoscope analyzes structures, one atom at a time, and "zooms out" as it digitally reconstructs the item of interest at a rate of millions of atoms per minute. This creates an unprecedented level of visibility and information at the atomic level.

Advances in nanoscale tools help us control and analyze matter more precisely, which in turn allows us to produce better tools. To summarize, the top-down path is designed and engineered with the following:

- Semiconductor industry adjacencies (with the benefits of market extensions and revenue along the way and the limitation of planar manufacturing techniques)
- Interfaces of scale inherited from the top

The Biological, Bottom-Up Path

In contrast to the top-down path, the biological bottom-up archetype is

- Grown via replication, evolution, and self-assembly in a 3-D, fluid medium
- Constrained at interfaces to the inorganic world
- Limited by gaps in learning and theory (in systems biology, complexity theory, and the pruning rules of emergence)
- Bootstrapped by a powerful preexisting hierarchy of interpreters of digital molecular code

To elaborate on this last point, a ribosome takes digital instructions in the form of mRNA and manufactures almost everything we care about in our bodies from a sequential concatenation of amino acids into proteins. The ribosome is a wonderful existence proof of the power and robustness of a molecular machine. It is roughly 20nm on a side and consists of only 99,000 atoms. Biological systems are replicating machines that parse molecular code (DNA) and a variety of feedback to grow macroscale beings. These highly evolved systems can be hijacked and reprogrammed to great effect.

So how does this help with the development of molecular electronics or nanotech manufacturing? The biological bootstrap provides a more immediate path to nanotech futures. Biology provides us with a library of prebuilt components and subsystems that can be repurposed and reused, and research in various labs is well under way in reengineering the information systems of biology.

For example, researchers at NASA's Ames Research Center are taking self-assembling heat shock proteins from thermophiles and genetically modifying them so that they will deposit a regular array of electrodes with a 17nm spacing. This could be useful for making patterned magnetic media in the disk drive industry or electrodes in a polymer solar cell.

At MIT, researchers are using accelerated artificial evolution to rapidly breed an Ml3 bacteriophage to infect bacteria in such a way that they bind and organize semiconducting materials with molecular precision.

At the Institute for Biological Energy Alternatives (IBEA), Craig Venter and Hamilton Smith are leading the Minimal Genome Project. They take *Mycoplasma genitalium* from the human urogenital tract and strip out 200 unnecessary genes, thereby creating the simplest organism that can self-replicate. Then they plan to layer new functionality onto this artificial genome, such as the ability to generate hydrogen from water using the sun's energy for photonic hydrolysis.

The limiting factor is our understanding of these complex systems, but our pace of learning has been compounding exponentially. We will learn more about genetics and the origins of disease in the next ten years than we have in all of human history. And for the minimal genome microbes, the possibility of understanding the entire proteome and metabolic pathways seems tantalizingly close to achievable. These simpler organisms have a simple "one gene, one protein" mapping and lack the nested loops of feedback that make the human genetic code so rich.

An Example: Hybrid Molecular Electronics

In the near term, a variety of companies are leveraging the power of organic self-assembly (bottom-up) and the market interface advantages of top-down design. The top-down substrate constrains the domain of self-assembly.

Based in Denver, ZettaCore builds molecular memories from energetically elegant molecules that are similar to chlorophyll. ZettaCore's synthetic organic porphyrin molecule self-assembles on exposed silicon. These molecules, called multiporphyrin nanostructures, can be oxidized and reduced (their electrons removed or replaced) in a way that is stable, reproducible, and reversible. In this way, the molecules can be used as a reliable storage medium for electronic devices.

Furthermore, the molecules can be engineered to store multiple bits of information and to maintain that information for relatively long periods before needing to be refreshed. Recall the water-drop-to-transistor-count comparison, and add to that the fact that these multiporphyrins have already demonstrated as many as eight stable digital states per molecule.

The technology has future potential to scale to 3-D circuits with minimal power dissipation, but initially it will enhance the weakest element of an otherwise standard 2-D memory chip. To end customers, the ZettaCore memory chip looks like a standard memory chip; nobody needs to know that it has "nano inside." The input/output pads, sense amps, row decoders, and wiring interconnect are produced via a standard semiconductor process. As a final manufacturing step, the molecules are splashed on the wafer, where they self-assemble in the predefined regions of exposed metal.

From a business perspective, this hybrid product design allows an immediate market entry because the memory chip defines a standard product feature set, and the molecular electronics manufacturing process need not change any of the prior manufacturing steps. Any interdependencies with the standard silicon manufacturing steps are also avoided, thanks to this late coupling; the fab can process wafers as it does now before spin-coating the molecules. In contrast, new materials for gate oxides or metal interconnects can have a number of effects on other processing steps, and these effects need to be tested. That introduces delay (as with copper interconnects).

Generalizing from the ZettaCore experience, the early revenue in molecular electronics will likely come from simple 1-D structures such as chemical sensors and self-assembled 2-D arrays on standard substrates, such as memory chips, sensor arrays, displays, CCDs for cameras, and solar cells.

IP and Business Model

Beyond product development time lines, the path to commercialization is dramatically impacted by the cost and scale of the manufacturing ramp. Partnerships with industry incumbents can be an accelerant or an albatross for market entry.

The strength of the IP protection for nanotech relates to the business models that can be safely pursued. For example, if the composition of matter patents afford the nanotech start-up the same degree of protection as for a biotech start-up, then a "biotech licensing model" may be possible in nanotech. A molecular electronics company could partner with a large semiconductor company for manufacturing, sales, and marketing, just as a biotech company partners with a big pharmaceutical partner for clinical trials, marketing, sales, and distribution. In both cases, the cost to the big partner is on the order of $100 million, and the start-up earns a royalty on future product sales.

Notice how the transaction costs and viability of this business model option pivot on the strength of IP protection. A software business, on the other end of the IP spectrum, would be very cautious about sharing its source code with Microsoft in the hopes of forming a partnership based on royalties.

Manufacturing partnerships are common in the semiconductor industry, with the "fabless" business model. This layering of the value chain separates the formerly integrated functions of product conceptualization, design, manufacturing, testing, and packaging. This has happened in the semiconductor industry because the capital cost of manufacturing is so large. The fabless model is a useful way for a small company with a good idea to bring its own product to market, but the company then must face the issue of gaining access to its market and funding the development of marketing, distribution, and sales.

Having looked at the molecular electronics example in some depth, we can now move up the abstraction ladder to aggregates, complex systems, and the potential to advance the capabilities of Moore's Law in software.

SYSTEMS, SOFTWARE, AND OTHER ABSTRACTIONS

Unlike memory chips, which have a regular array of elements, processors and logic chips are limited by the rat's nest of wires that span the chip on multiple layers. The bottleneck in logic chip design is not raw numbers of transistors but the lack of a design approach that can use all that capability in a timely fashion. For a solution, several next-generation processor companies have redesigned "systems on silicon" with a distributed computing bent; wiring bottlenecks are localized, and chip designers can be more productive by using a high-level programming language instead of wiring diagrams and logic gates. Chip design benefits from the abstraction hierarchy of computer science.

Compared with the relentless march of Moore's Law, the cognitive capability of humans is relatively fixed. We have relied on the compounding power of our tools to achieve exponential progress. To take advantage of accelerating hardware power, we must further develop layers of abstraction in software to manage the underlying complexity. For the next thousandfold improvement in computing, the imperative will shift to the growth of distributed complex systems. Our inspiration will likely come from biology.

As we race to interpret the now complete map of the human genome and embark upon deciphering the proteome, the accelerating pace of learning is not only opening doors to the better diagnosis and treatment of disease but is also a source of inspiration for much more powerful models of computer programming and complex systems development.

The Biological Muse

Many of the interesting software challenges relate to growing complex systems or have other biological metaphors as inspiration. Some of the interesting areas include Biomimetics, Artificial Evolution, Genetic Algorithms, A-life, Emergence, IBM's Autonomic Computing initiative, Viral Marketing, Mesh, Hives, Neural Networks, and the Subsumption architecture in robotics. The Santa Fe Institute just launched a BioComp research initiative.

In short, biology inspires IT, and IT drives biology. But how inspirational are the information systems of biology? If we took your entire genetic code—the entire biological program that resulted in your cells, organs, body, and mind—and burned it into a CD, it would be smaller than Microsoft Office. Just as images and text can be stored digitally, two digital bits can encode the four DNA bases (A, T, C, and G), resulting in a 750MB file that can be compressed for the preponderance of structural filler in the DNA chain.

If, as many scientists believe, most of the human genome consists of vestigial evolutionary remnants that serve no useful purpose, then we could compress it to 60MB of concentrated information. Having recently reinstalled Office, I am humbled by the comparison between its relatively simple capabilities and the wonder of human life. Much of the power in bioprocessing comes from the use of nonlinear fuzzy logic and feedback in the electrical, physical, and chemical domains.

For example, in a fetus, the initial interneuronal connections, or "wiring," of the brain follow chemical gradients. The massive number of interneuron connections in an adult brain could not be simply encoded in our DNA, even if the entire DNA sequence were dedicated to this one task. Your brain has on the order of 100 trillion synaptic connections between 60 billion neurons.

This highly complex system is not "installed," like Microsoft Office, from your DNA. Rather, it is grown, first through widespread connectivity sprouting from "static storms" of positive electrochemical feedback, and then through the pruning of many underused connections through continuous usage-based feedback. In fact, human brains hit their peak at the age of two to three years, with a quadrillion synaptic connections and twice the energy burn of an adult brain.

The brain has already served as an inspirational model for artificial intelligence (AI) programmers. The neural network approach to AI involves the fully interconnected wiring of nodes, followed by the iterative adjustment of the strength of these connections through numerous training exercises and the back-propagation of feedback through the system.

Moving beyond rule-based AI systems, these artificial neural networks are capable of many humanlike tasks, such as speech and visual pattern recognition, with a tolerance for noise and other errors. These systems shine precisely in the areas where traditional programming approaches fail.

The coding efficiency of our DNA extends beyond the leverage of numerous feedback loops to the complex interactions between genes. The regulatory genes produce proteins that respond to external or internal signals to regulate the activity of previously produced proteins or other genes. The result is a complex mesh of direct and indirect controls.

This nested complexity implies that genetic reengineering can be a very tricky endeavor if we have partial knowledge about the systemwide side effects of tweaking any one gene. For example, recent experiments show that genetically enhanced memory comes at the expense of enhanced sensitivity to pain.

By analogy, our genetic code is a dense network of nested hyperlinks, much like the evolving Web. Computer programmers already tap into the power and efficiency of indirect pointers and recursive loops. More recently, biological systems have inspired research in evolutionary programming, where computer programs are competitively grown in a simulated environment of natural selection and mutation. These efforts could transcend the local optimization inherent in natural evolution.

But therein lies great complexity. We have little experience with the long-term effects of the artificial evolution of complex systems. Early subsystem work can be deterministic of emergent and higher-level capabilities, as with the neuron. (Witness the Cambrian explosion of structural complexity and intelligence in biological systems once the neuron enabled something other than nearest-neighbor intercellular communication. Prior to the neuron, most multicellular organisms were small blobs.)

Recent breakthroughs in robotics were inspired by the "subsumption architecture" of biological evolution—using a layered approach to assemble reactive rules into complete control systems from the bottom up. The low-level reflexes are developed early and remain unchanged as complexity builds. Early subsystem work in any subsumptive system can have profound effects on its higher-order constructs. We may not have a predictive model of these downstream effects as we are developing the architectural equivalent of the neuron.

The Web is the first distributed experiment in biological growth in technological systems. Peer-to-peer software development and the rise of low-cost Web-connected embedded systems raise the possibility that complex artificial systems will arise on the Internet, rather than on one programmer's desktop. We already use biological metaphors, such as "viral" marketing, to describe the network economy.

Nanotech Accelerants: Quantum Simulation and High-Throughput Experimentation

We have discussed the migration of the lab sciences to the innovation cycles of the information sciences and Moore's Law. Advances in multiscale molecular modeling are helping some companies design complex molecular systems in silicon. But the quantum effects that underlie the unique properties of nanoscale systems are a double-edged sword. Although scientists have known for nearly 100 years how to write down the equations that an engineer needs to solve in order to understand any quantum system, no computer has ever been built that is powerful enough to solve them. Even today's most powerful supercomputers choke on systems bigger than a single water molecule.

This means that the behavior of nanoscale systems can be reliably studied only by empirical methods—building something in a lab and then poking and prodding it to see what happens.

This observation is distressing on several counts. We would like to design and visualize nanoscale products in the tradition of mechanical engineering, using CAD-like (Computer Aided Design) programs. Unfortunately this future can never be accurately realized using traditional computer architectures. The structures of interest to nanoscale scientists present intractable computational challenges to traditional computers.

The shortfall in our ability to use computers to shorten and reduce the cost of the design cycles of nanoscale products has serious business ramifications. If the development of all nanoscale products fundamentally requires long R&D cycles and significant investment, the nascent nanotechnology industry will face many of the difficulties that the biotechnology industry faces, without having a parallel to the pharmaceutical industry to shepherd products to market.

In a wonderful turn of poetic elegance, quantum mechanics itself turns out to be the solution to this quandary. Machines known as quantum computers, built to harness some simple properties of quantum systems, can perform accurate simulations of any nanoscale system of comparable complexity. The type of simulation conducted by a quantum computer results in an exact prediction of how a system will behave in nature—something that is literally impossible for any traditional computer, no matter how powerful.

Once quantum computers become available, engineers working at the nanoscale will be able to use them to model and design nanoscale systems, just as today's aerospace engineers model and design airplanes—completely virtually—with no wind tunnels (or their chemical analogs).

This may seem strange, but really it's not. Think of it this way: Conventional computers are really good at modeling conventional (that is, nonquantum) stuff, such as automobiles and airplanes. Quantum computers are really good at modeling quantum stuff. Each type of computer speaks a different language.

One of our companies is building a quantum computer using aluminum-based circuits. The company projects that by 2008 it will be building thumbnail-sized chips that will have more computing power than the aggregate total of all computers on the planet today and ever built in history, when applied to simulating the behavior and predicting the properties of nanoscale systems—thereby highlighting the vast difference between the capabilities of quantum computers and those of conventional computers. This would be of great value to the development of the nanotechnology industry. And it's a jaw-dropping claim. Professor David Deutsch of Oxford summarized it this way: "Quantum computers have the potential to solve problems that would take a classical computer longer than the age of the universe."

Although any physical experiment can be regarded as a complex computation, we will need quantum computers to transcend Moore's Law into the quantum domain to make this equivalence realizable. In the meantime, scientists will perform experiments. Until recently, the methods used for the discovery of new functional materials differed little from those used by scientists and engineers a hundred years ago. It was very much a manual, skilled-labor-intensive process. One sample was prepared from millions of possibilities, then it was tested, and then the results were recorded and the process repeated. Discoveries routinely took years.

Companies like Affymetrix, Intematix, and Symyx have made major improvements in a new methodology: high-throughput experimentation. For example, Intematix performs high-throughput synthesis and screening of materials to produce and characterize these materials for a wide range of technology applications. This technology platform enables the company to discover compound materials solutions more than 100 times as fast as with conventional methods. Initial materials have been developed that have applications in wireless communications, fuel cells, batteries, x-ray imaging, semiconductors, LEDs, and phosphors.

Combinatorial materials discovery replaces the old traditional method by simultaneously generating a multitude of combinations—possibly all feasible combinations—of a set of raw materials. This "materials library" contains all combinations of a set of materials, and they can be quickly tested in parallel by automated methods similar to those used in combinatorial chemistry and the pharmaceutical industry. What used to take years to develop now takes only months.

TIME LINE

Given our discussion of the various factors affecting the commercialization of nanotechnologies, how do we see them sequencing?

Early Revenue

- Tools and bulk materials (powders, composites). Several revenue-stage and public companies already exist in this category.
- One-dimensional chemical and biological sensors, out-of-body medical sensors and diagnostics.
- Larger MEMS-scale devices.

Medium Term

- Two-dimensional nanoelectronics: memory, displays, solar cells.
- Hierarchically structured nanomaterials and bio–nano assembly.
- Efficient energy storage and conversion.
- Passive drug delivery and diagnostics, improved implantable medical devices.

Long Term

- Three-dimensional nanoelectronics.
- Nanomedicine, therapeutics, and artificial chromosomes.
- Quantum computers used in small-molecule design.
- Machine-phase manufacturing.

The safest long-term prediction is that the most important nanotech developments will be the unforeseen opportunities, something that we cannot predict.

In the long term, nanotechnology research might ultimately enable miniaturization to a magnitude never before seen and might restructure and digitize the basis of manufacturing—such that matter becomes code. As with the digitization of music, the importance of this development is not only in the fidelity of reproduction but also in the decoupling of content from distribution. Once a product is digitized, new opportunities arise, such as online music swapping—transforming an industry.

With replicating molecular machines, physical production itself migrates to the rapid innovation cycle of information technology. With physical goods, the basis of manufacturing governs inventory planning and logistics, and the optimal distribution and retail supply chain has undergone little radical change for many decades. Flexible, low-cost manufacturing near the point of consumption could transform the physical goods economy and even change our notion of ownership—especially for infrequently used objects.

These are profound changes in the manufacturing of everything, changes that will ripple through the fabric of society. The science futurists have pondered the implications of being able to manufacture anything for $1 per pound. And as some of these technologies couple tightly to our biology, it will bring into question the nature and extensibility of our humanity.

THE ETHICAL DEBATE: GENES, MEMES, AND DIGITAL EXPRESSION

These changes may not be welcomed smoothly, especially with regard to reengineering the human germ line. At the societal level, we will likely try to curtail the evolution of evolvability and "genetic free speech." Larry Lessig predicts that we will recapitulate the 200-year debate about the first amendment to the Constitution. Pressures to curtail free genetic expression will focus on the dangers of "bad speech"; others will argue that good genetic expression will crowd out the bad, as it did with mimetic evolution (in the scientific method and the free exchange of ideas). Artificial chromosomes with adult-trigger events can decouple the agency debate about parental control. And, with a touch of irony, China may lead the charge.

We subconsciously cling to the selfish notion that humanity is the end point of evolution. In the debates about machine intelligence and genetic enhancements, there is a common and deeply rooted fear of being surpassed—in our lifetime. When the question is framed as a matter of parenthood (would you want your great-grandchild to be smarter and healthier than you?), the emotion often shifts from a selfish sense of supremacy to a universal human search for symbolic immortality.

CONCLUSION

Although predicting the future is becoming more difficult with each passing year, we should expect an accelerating pace of technological change. We conclude that nanotechnology is the next great technology wave and the next phase of Moore's Law. Nanotech innovations enable an array of disruptive businesses, driven by entrepreneurship, that were not possible before.

Much of our future context will be defined by the accelerating proliferation of information technology—as it innervates society and begins to subsume matter into code. It will be a period of exponential growth in the impact of the learning-doing cycle, where the power of biology, IT, and nanotech compounds the advances in each formerly discrete domain.

At DFJ, we conclude that it is a great time to invest in start-ups. As in evolution and the Cambrian explosion, many will become extinct. But some will change the world. So we pursue the strategy of a diversified portfolio, or, in other words, we try to make a broad bet on mammals.

Chapter 5

Investment in Nanotechnology

Investment decisions will be a major force in shaping how and where nanotechnology develops. Nanotechnology investments will largely be handled through partnership and technology licensing between companies rather than by young companies experiencing explosive growth.

The perspective of venture capitalists (VCs) is an excellent one for instruction, even though only a small percentage (less than 5 percent) of companies fit the profile of a company that merits venture capital investment. Serving to quarterback the technology economy through investment decisions, a VC makes investments in those companies that, although they are risky, hold the most promise for the future. This chapter outlines both venture and public funding decisions.

VENTURE CAPITAL INVESTING

Daniel V. Leff

Venture capital is money that is typically invested in young, unproven companies with the potential to develop into multibillion-dollar industry leaders, and it has been an increasingly important source of funds for high-technology start-up companies in the last several years. Venture capitalists are the agents that provide these financial resources as well as business guidance in exchange for ownership in a new business venture. VCs typically hope to garner returns in excess of 30–50 percent per year on their investments. They expect to do so over a four- to seven-year time horizon, which is the period of time, on average, that it takes a start-up company to reach a liquidity event (a merger, acquisition, or initial public offering).

Very few high-tech start-up companies are attractive candidates for VC investment. This is especially true for nanotechnology start-ups, because the commercialization of nanoscience is still in its nascent stages. Companies that are appropriate for VC investment generally have some combination of the following five characteristics: (1) an innovative (or disruptive) product idea based on defensible intellectual property that gives the company a sustainable competitive advantage; (2) a large and growing market opportunity that is greater than $1 billion and is growing at more than 20–30 percent per year; (3) reasonable time to market (one to three years) for the first product to be introduced; (4) a strong management team of seasoned executives; and (5) early customers and relationships with strategic partners, with a strong likelihood of significant revenue.

An early-stage start-up company rarely possesses all of these characteristics and often does not need to in order to attract venture financing. Indeed, early-stage start-ups are often funded without complete management teams, strategic partners, or customers. Absent these characteristics, however, there should be, at a minimum, a passionate, visionary entrepreneur who helped develop the core technology and wants to play an integral role in building the company.

Nanotechnology Venture Capital Investment

Nanotechnology is not a single market but rather a set of enabling (and potentially groundbreaking) technologies that can be applied to solve high-value problems in almost every industry. This includes industries as disparate as telecommunications, biotechnology, microelectronics, textiles, and energy. Many investors refer to nanotechnology investing as if it were its own investment category, because nanotechnology can add unique and specific value to a product that results in greatly enhanced performance attributes or cost advantages (or both). But customers purchasing nanotechnology products are buying these products, not because they are based on nanotechnology, but because they are characterized by specific performance enhancements, reduced costs, or both.

Almost every product application of nanotechnology is based either on a material characterized by nanoscale dimensions or on a process technology conducted at the nanometer scale. Nanomaterials possess unique properties—including optical, electronic, magnetic, physical, and chemical reactivity properties—that, when harnessed appropriately, can lead to entirely new, high-performance technologies and products. Changing a material's size, rather than its chemical composition, enables the control of that material's fundamental properties.

Nanotechnology Start-up Companies

Nanotechnology start-up companies should not expect to defy fundamental business principles, as did the Internet companies of the mid- to late 1990s, if only for a brief period. Nanotechnology companies should expect to be measured by standard metrics and to confront the same industry dynamics and fundamental business issues (for example, personnel choices, sales strategy, high-volume manufacturing, efficient allocation of capital, marketing, execution of their business model, time-to-market challenges, and so on) that face the other companies in their relevant industry category.

Certain key characteristics often differentiate nanotechnology start-up companies. They possess a technology platform with a body of intellectual property and a team of scientists, but no formal business plan, product strategy, well-defined market opportunity, or management team. Second, they are founded by (or are associated with) leading researchers at top-tier academic institutions. They employ a financing approach that highly leverages equity financing with the application of grant funding, and they need to have a more scientifically diverse workforce than other start-up companies.

It is common for these companies to employ chemists, physicists, engineers, biologists, computer scientists, and materials scientists because of the interdisciplinary nature of nanotechnology and the unique skills and knowledge that are required for product commercialization. Moreover, nanotech companies tend to sign up development partners (usually larger, more established companies) early in their maturation to provide technology validation and additional resources in the form of development funds, access to technology, sales and distribution channels, and manufacturing expertise.

Nanotechnology start-up companies can best be classified into six primary categories: nanomaterials and nanomaterials processing; nanobiotechnology; nanosoftware; nanophotonics; nanoelectronics, and nanoinstrumentation. Many companies in the nanomaterials category are developing methods and processes to manufacture a range of nanomaterials in large quantities as well as developing techniques to functionalize, solubilize, and integrate these materials into unique formulations. A variety of nanomaterials will ultimately be integrated into a host of end products (several are on the market) that will provide unique properties, such as scratch resistance, increased stiffness and strength, reduced friction and wear, greater electrical and thermal conductivity, and so on.

The three areas that have received the most funding based on dollars invested are nanoelectronics, nanophotonics, and nanoinstrumentation. However, in terms of the absolute number of companies that have been funded, nanomaterials companies are the clear leader.

Nanobiotechnology is the application of nanotechnology to biological systems. Applications exist in all of the traditional areas of biotechnology, such as therapeutics discovery and production, drug-delivery systems technologies, diagnostics, and so on. Incorporating nanotechnology into biotechnology will lead to the enhanced ability to label, detect, and study biological systems (such as genes, proteins, DNA fragments, single molecules, and so on) with great precision as well as to develop unique drug targets and therapies.

Nanoelectronics is based upon individual or ordered assemblies of nanometer-scale device components. These building blocks could lead to devices with significant cost advantages and performance attributes, such as extremely low power operation (~nanoWatt), ultra-high device densities (~1 trillion elements/cm^2), and blazing speed (~1 Terahertz switching rates). In addition, the possibility exists of enabling a new class of devices with unique functionality. Examples include, but are not limited to, multistate logic elements; high-quantum-efficiency, low-power, tunable, multicolor light-emitting diodes (LEDs); low-power, high-density nonvolatile random access memory (RAM); quantum dot-based lasers; universal analyte sensors; low-impedance, high-speed interconnects, and so on.

Nanophotonics companies are developing highly integrated, subwavelength optical communications components using a combination of proprietary nanomaterials and nanotech manufacturing technologies, along with standard complementary metal oxide semiconductor (CMOS) processing. This provides for the low-cost integration of electronic and photonic components on a single chip. Products in this category include low-cost, high-performance devices for high-speed optical communications, such as wavelength converters, tunable filters, polarization combiners, reconfigurable optical add/drop multiplexers (ROADMs), optical transceivers, and so on.

Nanoinstrumentation is based on tools that manipulate, image, chemically profile, and write matter on a nanometer-length scale (far less than 100nm). These tools include the well-known microscopy techniques such as transmission electron microscopy (TEM), scanning electron microscopy (SEM), and atomic force microscopy (AFM), as well as newer techniques such as dip-pen nanolithography (DPN), nanoimprint lithography (NIL), and atom probe microscopes for elucidating three-dimensional atomic composition and structure of solid materials and thin films. These are the basic tools that enable scientists and engineers to perform nanoscale science and to develop nanotechnology products.

Nanosoftware is based on modeling and simulation tools for research in advanced materials (cheminformatics) and the design, development, and testing of drugs in the biotechnology industry (bioinformatics). This category also includes electronic and photonic architecture, structure, and device modeling tools such as specific incarnations of electronic design automation (EDA) software or quantum simulations, and so on. In addition, one might further include proprietary software packages developed to operate nanoinstrumentation-based tools or interpret data collected from such instruments.

PUBLIC MARKETS AND NANOTECHNOLOGY COMPANIES

R. Douglas Moffat

Historically, public equity markets have provided capital for rapidly expanding firms having established products and seeking growth capital. Periodically, new technology or corporate growth models, combined with unusually heavy money flows into the stock market, fuel speculative demand for shares in new companies. Biotechnology investing has run in such cycles for more than 20 years. The Internet boom of the late 1990s reached unprecedented levels of irrational expectations and speculation. Other examples include the fuel cell boom of 2000–2001.

The public market's appetite for initial public offerings (IPOs) in a sector also is heavily influenced by the business model characteristics and the track record of the model for success. Biotech has achieved success in part because of the appetite for these firms by big pharmaceutical firms. Software stocks have proven to be fast growers without heavy capital investment.

Nanotech probably will be a big hit on Wall Street, but the timing will depend on progress achieved in moving products closer to market acceptance. Many of the nanoscience-enabled products being commercialized now are coming out of large companies. Examples include nanotube-based plasma televisions and personal care products. A limited number of smaller firms are introducing nanotech products in the short term. Most companies, however, are still refining the science behind paradigm-shifting technologies having massive potential. Commercialization issues include interfacing nanodevices with the macro environment, scalable manufacturing, and, in the health-care world, long FDA approval cycles.

Wall Street investors typically have preferred focused business models concentrated on growth from a narrowly defined technology or product group. Management focus historically has produced better execution and shareholder returns.

At this stage of nanotechnology development, however, intellectual property platforms based on broad patents (often coming from academia) are the main assets behind many companies. The applicability of this IP could cut across many markets and applications. Some firms have amassed broad IP by taking a portfolio approach to early-stage commercialization, an approach most stock investors do not favor. Such diversification, however, makes sense not only from a scientific point of view but also to lessen risks associated with potential patent litigation. The patent landscape in nanotech might be likened to the gold rush days, with overlapping claims.

Nanotechnology is different from other tech waves. First, the technology is often paradigm shifting, either creating new markets or providing quantum improvement in performance at a low cost. The enabling science probably is applicable to a wide variety of applications. In time, stock market investors may come to appreciate the power of a new nanotech business model, one with core IP at its center and with the prospects to spin off many companies with varied new products. The evolution of acceptable nanotech business models in public markets will depend in part on VC investors' willingness to extend funding horizons to allow firms to develop products.

There is significant buzz on Wall Street around nanotechnology. Leading Wall Street firms are beginning to commit resources to research and fund nanotechnology. A favorable environment is emerging for a successful nanotech début on the street.

Since the Internet bubble deflation in 2000, public equity markets have taken on a more risk-averse character. IPO investors have preferred to fund companies with established products, revenues, and profits as well as large companies restructured by private equity firms. A limited number of nanotechnology-enabled firms have been able to tap public equity markets. Public equity access likely will improve as nanotechnology firms move closer to the introduction of novel products having a clear path to revenue and profits. Equity issuance by nanotech firms likely will grow slowly over the next five years, gathering potentially explosive momentum thereafter.

Chapter 6

The Role of the U.S. Government in Nanoscale Science and Technology

Geoffrey M. Holdridge

One of the cornerstone principles of the United States, and of the free market system that has helped to make the United States one of the world's richest countries, is that the central government should play only those roles that are both critical to the nation's welfare and that private citizens, private industry, or state and local governments are either unable or unwilling to undertake. One role for the federal government that was envisioned from the start of the Republic by the framers of the U.S. Constitution was to "promote the Progress of Science and useful Arts."[1] This role was originally intended only for the protection of intellectual property and was the basis for the establishment of the U.S. Patent and Trademark Office (USPTO).

However, as science and technology have become ever more critical to the fate of nations and the well-being of their citizens, the U.S. government has taken on additional corollaries to this role outlined in the Constitution, including funding research and development, education, and critical research infrastructure, as well as measures to more effectively promote the transfer of "Science" into "useful Arts" for exploitation by private industry. This chapter outlines and explains the sometimes unique roles that the U.S. federal government plays in the development of nanoscale science and technology. These roles can be expected to be similar in other countries as well.

The most fundamental government role in nanoscale science and technology, or nanotechnology, is to support research and development—long-term basic research as well as development of applications relevant to specific national priorities (such as defense). Another key government role is developing educational resources, a skilled workforce, and the supporting infrastructure and tools (for example, laboratory facilities and instrumentation) needed to advance nanotechnology. The government has an important role to play in facilitating transfer of new technologies developed in university and government laboratories to the private sector, where the real benefits of nanotechnology (such as new and improved products, jobs, and sustainable economic growth) can be pursued.

Finally, the government has a responsibility for encouraging the responsible development of nanotechnology, by supporting research on the potential impacts (both positive and negative) of nanotechnologies and by creating a regulatory environment that protects public health and safety while providing a clear path in which industry can pursue development of novel products and services based on nanotechnology.

These roles are not unique to nanotechnology but rather reflect the roles that the U.S. government and many other governments around the world have played in the development and exploitation of science and technology at least since World War II. What is perhaps unique about the case of nanoscale science and technology is its accelerated pace of development, the broad spectrum of its potential benefits, and hence its potential to have a rapid and revolutionary impact on industry and society.

Another unique aspect is the extent to which progress in nanotechnology stems from a confluence of developments in different fields of science and technology (for example, biology and physics), presenting new challenges and opportunities for researchers and educators working across traditional disciplinary boundaries. Government plays an important role in helping to foster interdisciplinary research and education in three ways: (1) by supporting the development of and making available facilities and research instruments that allow researchers from all disciplines to visualize and manipulate matter at the nanometer scale, (2) by establishing research centers and groups with the explicit goal of bringing together researchers from previously divergent fields, and (3) by promulgating policies to encourage the education of a new generation of multidisciplinary researchers.

THE NATIONAL NANOTECHNOLOGY INITIATIVE AND THE 21ST CENTURY NANOTECHNOLOGY RESEARCH AND DEVELOPMENT ACT

The U.S. National Nanotechnology Initiative (NNI) was proposed in 2000 following the realization that the ability to understand and control matter on the nanometer scale was creating a revolution in science, technology, and industry, and with the vision of accelerating and guiding that revolution for the benefit of society. Because of the breadth of nanotechnology's disciplinary roots and of its potential applications, it relates to the missions of an extraordinary number of U.S. government agencies.

At its outset the NNI involved eight agencies. As of this writing, eleven agencies have specific funding under the NNI, and an additional eleven participate as partners and in-kind contributors because the development of nanotechnology is relevant to their missions or regulatory roles. To coordinate this broad interagency effort, the President's National Science and Technology Council has established a Subcommittee on Nanoscale Science, Engineering, and Technology (NSET), with representatives from all twenty-two NNI participating agencies. NSET in turn has established a National Nanotechnology Coordination Office (NNCO), with a small full-time staff to provide technical and administrative support to NSET, to serve as the point of contact on federal nanotechnology activities, to conduct public outreach, and to promote technology transfer.

Another somewhat extraordinary aspect of nanotechnology is that it is the subject of an act of Congress. Of the thousands of bills that are proposed in Congress each year, few are ever signed into law, and even fewer of those (other than routine annual appropriations bills) concern science and technology. But in 2003 Congress passed the 21st Century Nanotechnology Research and Development Act, which was signed into law by President George W. Bush on December 3, 2003, as Public Law 108-153.[2] This act calls for the establishment of a National Nanotechnology Program, with the following mission:

> (1) establish the goals, priorities, and metrics for evaluation for Federal nanotechnology research, development, and other activities; (2) invest in Federal research and development programs in nanotechnology and related sciences to achieve those goals; and (3) provide for interagency coordination of Federal nanotechnology research, development, and other activities undertaken pursuant to the Program.

The NNI continues as the primary means by which the Executive Branch implements the National Nanotechnology Program, and for carrying out the many roles that it stipulates for the federal government in promoting the development and commercialization of nanotechnology.

In December 2004, in keeping with the provisions of PL 108-153, the NNI released a Strategic Plan that sets out its vision, goals, and strategies for achieving those goals.[3] The NNI's goals closely align with the government roles just discussed. The following sections briefly review each of those roles.

RESEARCH AND DEVELOPMENT

Prior to World War II the U.S. government played only a minor role in R&D funding. Private foundations, corporations, and internal university funds provided the bulk of support. The U.S. military funded some advanced development of technologies for deployment in Army and Navy equipment, but it did not provide significant support for basic research. The National Advisory Committee for Aeronautics (NACA, precursor to NASA) played an important role in developing and testing new airfoil designs in the 1930s. The Department of Agriculture, through the Hatch Act of 1887, supported agricultural research at the land grant colleges. Federal support for basic research in the health sciences, and hence in biology, dates to the latter part of the nineteenth century but began as a major funding activity in the 1930s with the establishment of the National Institutes of Health. Except for these and a few other isolated examples, however, there was no systematic federal support for scientific research.

The obvious importance of science and technology to the war effort during World War II led to a new role for the U.S. government in supporting not only applied research for immediate military applications but also basic research. The August 1939 letter from Albert Einstein to President Roosevelt ultimately led to an understanding during and after WWII of the need for federal support for basic research in fields such as nuclear physics. The federal government has been the mainstay of funding for basic research in the physical sciences and engineering ever since, through agencies such as the Office of Naval Research, the Atomic Energy Commission, and the National Science Foundation, all of which trace their origins to the aftermath of World War II and the early days of the cold war.

Federal support for R&D increased steadily through the 1950s and early 1960s, including considerable funding for basic research in particle physics, electronics, and materials science needed for the space program and the military. This research in turn led to major benefits for the civilian economy, including civilian nuclear power, inexpensive and ubiquitous integrated circuits, the widespread use of computers in the banking industry, and dramatic advances in air travel. All these advances stemmed in part from increased understanding and control of the properties of matter at the atomic and molecular scale.

Federal R&D funding leveled off and even declined somewhat in real dollars starting in the late 1960s, but it increased from the late 1970s through the mid-1980s, with increases in federal support for energy and military research.[4] The National Institutes of Health also received sustained funding increases, notably with the advent of Richard Nixon's war on cancer and more recently with the bipartisan pledge to double the overall NIH budget in the first few years of the twenty-first century.

Throughout this period the federal investment in research continued to return significant dividends for military security, the civilian economy, and health care—with advances in military electronics that dramatically improved the performance of U.S. forces, novel treatments for many diseases, and the development of the Internet. The U.S. standard of living and life expectancy grew dramatically during the entire post–World War II period, fueled to a large extent by advances in science and technology funded by the U.S. government.

In fact, basic research in particle physics and materials science conducted in the national laboratory system during the latter half of the twentieth century greatly enhanced our understanding of the properties of matter on the atomic and molecular scales, developing along the way novel methods of measuring, visualizing, and controlling these properties. Research funded by the National Institutes of Health improved our understanding of the basic operating principles of life itself. Much of this basic and applied research funded by the federal government in the latter half of the twentieth century provided the foundation on which the nanotechnology revolution of the twenty-first century will be constructed.

The Federal Role in Nanoscale Research and Development: Basic and Applied

Although political parties have debated the role of the federal government in applied research over the last several decades, there has been largely unanimous support across a wide expanse of the political spectrum for basic research funding. That's because private corporations cannot justify large investments in basic science that, as important as they are to our future, may take decades to provide reasonable returns on the investments, and it may be difficult for private corporations to capture such returns. In recent years, this federal role has been even more important as large corporations such as AT&T and RCA have scaled down or eliminated the investments they had been making in basic research during the mid-twentieth century. NNI participating agencies such as the National Science Foundation (NSF) and the Office of Science at the Department of Energy (DOE) provide a key role in supporting basic nanoscale science and engineering research.

The appropriate federal role in supporting applied research has been debated publicly in recent years. The mission agencies—such as the Department of Defense (DoD), the National Aeronautics and Space Administration (NASA), and the National Institutes of Health (NIH)—have an agreed role in supporting applied research and even product development that will directly contribute to their respective missions. These agencies have identified numerous opportunities where applied research in nanotechnology can make important contributions to their missions. Examples include DoD support for development of nanotechnology-enabled sensors to detect chemical, biological, radiological, and explosives agents; NASA support for development of flexible, lightweight, and superstrong materials for future spacecraft; and NIH support for development of nanostructures that can be used to selectively target tumor cells with either radiation or chemotherapy.

Research at the intersection of traditional scientific and engineering disciplines is a requirement if we are to make rapid progress in nanoscale science, technology, and applications. The agencies participating in the NNI encourage such interdisciplinary research through several mechanisms, including funding of small multidisciplinary teams as well as large research centers involving faculty from many different disciplinary departments, and ultimately promoting the development of a new generation of young researchers not bound by the traditional disciplinary divisions.

EDUCATION AND WORKFORCE DEVELOPMENT

Support for the education of the next generation of nanoscale science and engineering researchers is key to continued progress in nanotechnology, and thus it represents another key federal role. The NNI provides support for hands-on training of undergraduates, graduate students, and postdoctoral researchers at universities, federal laboratories, and other research institutions. Participating agencies award funding directly to students for fellowships and traineeships.

NNI agencies also encourage the development of new curricula at various educational levels, curricula designed to nurture the new generation of interdisciplinary researchers. A variety of programs are targeted at bringing the concepts of nanoscale science and technology into classrooms for high school and even middle school students. The exciting opportunities offered by nanotechnology are providing a new incentive for students to take an early and continuing interest in pursuing careers in science and technology generally.

Equally important is the need for a workforce familiar with the concepts of nanoscale science and technology and equipped with the skills that will be needed in the nanotechnology-enabled industries of the future. NNI participating agencies support development of educational programs that have the objective of training technicians to meet the growing demand from industry, as nanotechnology applications move into more products and services. Such programs are targeted to community colleges and other institutions that emphasize development of job-related skills and knowledge.

It also is vital for the general population to be informed about nanotechnology. A profusion of science-fiction literature and video, as well as various Web sites, promulgates information and conjecture that may mislead some people into thinking of nanotechnology as either a panacea or a reason for panic. It is, of course, impossible to accurately foretell the future, especially to predict the long-term implications of nanotechnology on society. But in the near term, nanotechnology will not solve all the world's problems, nor will it dissolve the human race into blobs of "gray goo."

The NNI supports various efforts to communicate to the general populace the real promise of nanotechnology through venues such as science museum exhibits and outreach from NNI-funded research centers to elementary schools. NNI also is working to provide information on the real hazards that some aspects of nanotechnology may pose (such as the possible inhalation of certain types of nanoparticles) and to explain what the federal government is doing to clarify and address those issues (for example, information on environmental health and safety research found on the NNI Web site).[5]

FACILITIES AND INSTRUMENTATION

Nanoscale science and technology R&D often requires the use of complex and expensive facilities and instrumentation. In fact, it was the development of tools such as the scanning tunneling microscope (STM) at IBM/Zurich in the 1980s that many experts point to as among the key events that enabled the dawn of this new era of nanoscale science and technology. Tools like the STM are what make it possible to "image, measure, model, and manipulate matter on the nanoscale."[6] Effective use of such tools also can require specialized facilities to house them, including dedicated buildings designed to eliminate the vibration, particulate contamination, noise, and temperature variations typical of conventional research facilities. Without this vital infrastructure it would be impossible to realize the true promise of nanotechnology.

The NNI provides access to such facilities and instrumentation through a variety of means. Some large NNI-funded research centers can include funding for such infrastructure in their budgets. However, breakthroughs in science and technology often come from unexpected quarters. In particular, some of the most creative work often is conducted by individual investigators or small teams, which may or may not be affiliated with large universities housing NNI-funded research centers. Even more importantly, for the fruits of basic research to be developed into useful products, researchers in industry also require access to the essential tools of nanotechnology R&D. Only a few very large corporations can afford to purchase such infrastructure for their own exclusive use, whereas small start-up high-technology companies are often the source of some of the most important commercial innovations, and they are well known to be the engine of new job creation in the U.S. economy.

Therefore, a critical role for the government is to provide access to nanotechnology facilities and instrumentation for a wide spectrum of researchers, both in academia and in industry. Accordingly, a large component of the NNI funding is dedicated to the development and operation of user facilities, open to any qualified researchers on a merit-review basis. Two of the most significant NNI user facility programs are (1) the five DOE Nanoscale Science Research Centers (NSRCs) now under construction, each with a specific focus and associated with existing large-scale research facilities and competencies at DOE laboratories across the country; and (2) the NSF's National Nanotechnology Infrastructure Network, with existing and new facilities available at thirteen universities.[7] NSF also supports a seven-university network of specialized user facilities for research on modeling and simulation of nanostructures and related systems.[8]

Another related government role is support for development of new types of instrumentation for nanoscale science and technology. Just as the development of the AFM and other probe microscopes opened up whole new worlds for research and application in recent years, future progress likely will be enabled by new instrumentation breakthroughs. The National Institute of Standards and Technology (NIST) plays a key role in conducting research on novel instrumentation. NSF also funds university research on new instrumentation concepts.

TECHNOLOGY TRANSFER

The NNI vision is that nanoscale science and technology will lead to "a revolution in technology and industry."[9] To realize that vision, the results of NNI-funded R&D must be effectively transferred to private industry, where technologies are incorporated into new and improved products and services. Although the government role stops short of directly commercializing nanotechnology per se—because this role is most appropriately carried out in the private sector—it is not surprising that technology transfer of NNI-funded research *to* the private sector is among the four principal goals of the NNI. This transfer can occur via various pathways, such as publication of the results of federally funded research, hiring by industry of recent graduates of

NNI-funded educational programs, and licensing of intellectual property resulting from federally funded research.

The traditional role of the USPTO is critical to the latter example. USPTO has undertaken a number of activities to enhance its ability to address the intellectual property issues associated with nanotechnology. Examples include training programs for its patent examiners, networking with its European and Japanese counterparts, and efforts to identify and classify nanotechnology-related patents, thus improving its ability to search nanotechnology-related prior art.

A primary aspect of all technology transfer is interaction among those who are performing R&D and those who manufacture and sell goods and services. NSET has established liaisons with various commercial sectors to promote the exchange of information on NNI research programs and industry needs that relate to nanotechnology. Such activities are under way with the semiconductor, electronics, and chemical industries and are in development with the biotechnology, aerospace, and automotive industries.

The NNI agencies also support meetings at which researchers from academia, government, and industry exchange information on results and possible applications. The NNI-funded user facilities help to foster interaction among industry, academic, and government researchers who use the facilities. The NSF-funded research centers include industry partners, allowing industry to communicate its needs at an early stage and increasing the likelihood that new ideas will ultimately be developed and commercialized. The government-wide Small Business Innovation Research (SBIR) and Small Business Technology Transfer (STTR) programs support the early-stage development of many nanotechnology-based small start-up companies, frequently spun off from NNI-supported university research.

The government also plays a key role in supporting standards development, another activity important to successful commercialization of nanotechnology. The American National Standards Institute (ANSI) established the Nanotechnology Standards Panel (NSP) in September 2004. The present director of the NNCO is the government co-chair of the NSP. The National Institute of Standards and Technology (NIST) also plays a central role for the government in standards development for nanotechnology.

RESPONSIBLE DEVELOPMENT

Given the breadth of the expected applications of nanotechnology and their likely revolutionary impact on industry and society, another important role for the Government is to support research and other activities that address the broad social implications of nanotechnology, including benefits and risks. Responsible development of nanotechnology comprises one of the four principal goals outlined in the 2004 NNI Strategic Plan, and it is also one of the principal components of the NNI research and education program. Activities supported by the NNI consistent with this role include support for research on the environmental, health, and safety impacts of nanotechnology, education-related activities, and research directed at identifying and quantifying the broad implications of nanotechnology for society, including social, economic, workforce, educational, ethical, and legal implications.

Perhaps the most pervasive concern that has been raised in the popular press about potential negative impacts of nanotechnology relates to its potential for adverse environmental, health, or safety implications. The NSET Subcommittee has formed a formal interagency working group to exchange information among regulatory and R&D agencies on this subject, including participation by several agencies within the Executive Office of the President.[10] This group also seeks to help prioritize and implement research required for the responsible development and oversight of nanotechnology, and to promote communication of information related to the responsible handling and use of nanotechnology-enabled products. Some industry groups have expressed the view that the government needs to provide a clear regulatory path for industry to follow for the safe and responsible commercialization of nanotechnology, as a key government role necessary to successful industrial development.

Existing Laws and Regulations and the Responsible Development of Nanotechnology

Some people have asked whether there is a need for new regulations or even statutes governing the development of nanotechnology to avoid any possible adverse consequences to the environment, public health, worker safety, or even economic stability. Although research on the environmental, health, safety, and societal impacts of nanotechnology is in its infancy, this may remain an open question. However, with the information available to date, it is difficult to imagine a scenario in which the existing legal and regulatory framework that has evolved over the past century to address such concerns would need to be changed significantly.

The following discussion outlines some of the existing laws, regulations, and procedures that are in place to protect public health and safety that might apply to nanotechnology. These should obviate the need for any major legal or regulatory changes.

Research Agencies

The U.S. government agencies that fund scientific research already have in place various policies to address ethical, health, and safety considerations related to the research they are funding or performing in their laboratories. Applicants for NSF research funding are required to certify in writing that they are in compliance with various NSF policies and relevant sections of the U.S. Code of Federal Regulations with respect to, for example, recombinant DNA, human subjects, and the use of animals in research. NSF has broad authority to suspend or terminate ongoing research grants if grantees fail to comply with grant regulations, or for "other reasonable cause."[11] NSF program officers conduct regular site visits to review ongoing research, where they have an opportunity to see whether there are any problems that might justify such actions. Other research agencies have comparable mechanisms.

Regulatory Agencies

For research and product development not funded by the government, other statutory and practical controls apply. The many existing statutes and regulations addressing commercial products also apply to nanotechnology-based products and materials. For example, the Consumer Products Safety Act of 1972 (administered by the Consumer Products Safety Commission) requires that manufacturers of consumer products ensure that their products are safe and holds them liable if they are not; CPSC can require the recall of products in cases where it has evidence that they are not safe.[12] The Occupational Safety and Health Act of 1970 (administered by the Occupational Safety and Health Administration, or OSHA, within the Department of Labor) provides for federal regulation of safety in the workplace, including public as well as private research laboratories.[13]

The National Institute of Occupational Safety and Health (NIOSH), within the Centers for Disease Control, conducts research and training to inform OSHA regulatory decisions.[14] NIOSH is leading an effort within the government to develop a set of recommended safe handling practices for nanomaterials, for research as well as commercial production facilities.

The Toxic Substances Control Act of 1976 (administered by the Environmental Protection Agency) regulates the manufacture, importation, and use of new and existing chemical substances.[15] Several other statutes administered by the EPA (such as the Clean Air Act and the Clean Water Act) also may come into play with respect to nanotechnology. The National Institute of Environmental Health Sciences, within the National Institutes of Health, also conducts health-related research that informs regulatory decisions by other agencies.[16] The Federal Food, Drug and Cosmetic Act (originally enacted as the Pure Food and Drug Act in 1906 and administered by the Food and Drug Administration within the Department of Health and Human Services) requires prior testing and review of pharmaceutical and medical products under strictly controlled conditions.[17]

Under most of these statutes, any adverse effects of new products or processes that are uncovered as a result of privately funded research must be reported to the government. These and many other statutes and accompanying regulations provide an ample basis for both criminal and civil legal action against any private organizations that produce or import products or services that may be deemed hazardous to the public. All of the agencies and institutes listed here are coordinating their nanotechnology-related activities through their participation in the NSET Subcommittee and its NEHI Working Group.

To clarify how these existing statutes and regulatory structures apply to nanotechnology, scientific as well as legal research efforts are under way. For example, NSET member agencies are working with universities, industry, and standards organizations to develop a clear system of nomenclature for classifying new nanomaterials. This nomenclature standard is needed to facilitate both appropriate regulation of nanotechnology and its commercial development. In another example, EPA will be holding a public hearing in 2005 to inform discussion of how the Toxic Substances Control Act should be applied to nanomaterials, starting with a voluntary pilot program.[18]

For nanotechnology R&D specifically, the 21st Century Nanotechnology Research and Development Act includes a number of additional measures to ensure adequate oversight of the interagency NNI activity. One of the stated purposes of the Act is to ensure that "ethical, legal, environmental, and other appropriate societal concerns" are properly addressed. Specific measures include oversight of the NNI by the President's National Science and Technology Council, a separate National Nanotechnology Advisory Panel, and a triennial review of the NNI by the National Academy of Sciences. All three of these oversight bodies are charged with addressing the responsible development of nanotechnology. The NNCO and the Office of Science and Technology Policy (part of the Executive Office of the President) also play coordinating and oversight roles, respectively, in the responsible management of the NNI research portfolio.

In short, within the United States, not only are the mechanisms for minimizing any misuses of nanotechnology in place, but also they are working to ensure the responsible and safe development of nanotechnology.

Geoffrey M. Holdridge is Vice President for Government Services, WTEC, Inc. (http://www.wtec.org), currently on assignment as

Policy Analyst at the U.S. National Nanotechnology Coordination Office (http://www.nano.gov/html/about/nnco.html). Any opinions, conclusions, or recommendations expressed in this material are those of the author and do not necessarily reflect the views of the United States government or of WTEC, Inc. Portions of text extracted from the National Nanotechnology Initiative Strategic Plan, http://www.nano.gov/NNI_Strategic_Plan_2004.pdf.

REFERENCES

1. U.S. Constitution, Article 1, Section 8. See http://www.house.gov/Constitution/Constitution.html.
2. 15 USC 7501, Page 117 STAT. 1923. See http://www.nano.gov/html/news/PresSignsNanoBill.htm for a press release and a pointer to the text of the law.
3. http://www.nano.gov/NNI_Strategic_Plan_2004.pdf.
4. http://www.aaas.org/spp/rd/rddis06b.pdf.
5. http://www.nano.gov/html/society/EHS.htm. See also the nanotechnology Web sites at the Food and Drug Administration (http://www.fda.gov/nanotechnology/), the Environmental Protection Agency (http://www.epa.gov/ordntrnt/ORD/accomplishments/nanotechnology.html), and the National Institute for Occupational Safety and Health (http://www.cdc.gov/niosh/topics/nanotech/).
6. NNI Strategic Plan, http://www.nano.gov/NNI_Strategic_Plan_2004.pdf, Preface, p. iii.
7. See http://nnin.org/.
8. Network for Computational Nanotechnology, http://ncn.purdue.edu/.
9. NNI Strategic Plan, http://www.nano.gov/NNI_Strategic_Plan_2004.pdf, Executive Summary, p. i.
10. Interagency Working Group on Nanotechnology Environmental and Health Implications (NEHI). See http://www.nano.gov/html/society/EHS.htm.
11. See the NSF Grant Policy Manual, http://www.nsf.gov/pubs/2002/nsf02151/, in particular Chapter VII, "Other Grant Requirements," and the sections in Chapter IX on research misconduct and on termination policies and procedures.
12. http://www.cpsc.gov.

13. http://www.osha.gov.
14. http://www.cdc.gov/niosh/.
15. http://www.epa.gov.
16. http://www.niehs.nih.gov.
17. http://www.fda.gov.
18. See U.S. Federal Register notice, volume 70, number 89 (May 10, 2005): 24574–24576 (http://www.gpoaccess.gov/fr/index.html).

Chapter 7

Overview of U.S. Academic Research

Julie Chen

A large portion of current research in nanotechnology is done by academic researchers, and understanding nanoscale behavior falls naturally within their sphere. The National Science Foundation (NSF) has the largest role in funding such research within the National Nanotechnology Initiative (NNI), with additional academic research being supported by the Department of Defense, Department of Energy, National Institutes of Health, National Aeronautics and Space Administration, Department of Agriculture, and Environmental Protection Agency.

The National Science Foundation is an independent federal agency that provides research funding for colleges and universities in many fields, such as mathematics, computer science, and the social sciences. By definition, nanotechnology represents a significant shift in the behavior of materials, devices, structures, and systems because of the nanoscale dimension. This fundamental research also extends to research in manufacturing science, because the assembly of nanoelements is a critical scientific barrier to moving nanotechnology toward application in commercial products.

Every university with a research program in science and engineering has some activity in nanotechnology, and a basic search of these programs is likely to provide a limited view. Recognizing that the majority of academic researchers rely on federal funding to pay for graduate students, equipment, and supplies, we can gain a better understanding of the multitude of research activities by understanding what types of research and education outcomes are most likely for the various federal funding mechanisms. As an example, this chapter provides basic information on some of the NSF funding structure, research focus areas, and resources for more information.

NATIONAL SCIENCE FOUNDATION FUNDING MECHANISMS

NSF has developed a multilevel approach to funding research in nanotechnology. The Nanoscale Science and Engineering (NSE) program has been funded for five years now. The funding structure ranges from small, one-year, exploratory grants to large five- to ten-year center awards. Each type of grant provides a different opportunity for researchers and a different level of technological maturity for companies that may be looking for new technology and partners in academia. NSF also has primary programs (not specifically related to nanotechnology) in the Directorates for Engineering, Math and Physical Sciences, and Biological Sciences; these programs fund substantial levels of research related to nanotechnology. Abstracts of funded research projects can be found at the NSF Web site (www.fastlane.nsf.gov).

Nanoscale Science and Engineering Centers (NSECs)

NSEC awards are typically for five years and are renewable for another five years. Although the actual funding for individual research projects within a center is not large, a key mission of these centers is to address interdisciplinary, systems-level research, as well as to build an infrastructure for both education and research. Each NSEC is also meant to be a focal point and portal for a specific research vision, leading advances in a specific theme while retaining knowledge about what others are doing in related areas. Currently fourteen NSECs have been established, with two others anticipated to be funded in the fiscal year 2005 competition (see Table 7–1).

Table 7–1 NSF-funded Nanoscale Science and Engineering Centers (NSECs)

Institution	*Center*	*Established*
Northwestern University	Integrated Nanopatterning and Detection Technologies (www.nsec.northwestern.edu)	Sept. 2001
Cornell University	Nanoscale Systems in Information Technologies (www.cns.cornell.edu)	Sept. 2001
Harvard University	Science of Nanoscale Systems and Their Device Applications (www.nsec.harvard.edu)	Sept. 2001

Institution	*Center*	*Established*
Columbia University	Electronic Transport in Molecular Nanostructures (www.cise.columbia.edu/nsec)	Sept. 2001
Rice University	Nanoscience in Biological and Environmental Engineering (www.cben.rice.edu)	Sept. 2001
Rensselaer Polytechnic Institute, University of Illinois	Directed Assembly of Nanostructures (www.rpi.edu/dept/nsec)	Sept. 2001
UCLA, UC Berkeley, UC San Diego, Stanford, UNC Charlotte	Center for Scalable and Integrated Nanomanufacturing (www.sinam.org)	Sept. 2003
University of Illinois Urbana-Champaign	Center for Nanoscale Chemical-Electrical-Mechanical-Manufacturing Systems (www.nano-cemms.uiuc.edu)	Sept. 2003
Northeastern University, University of Massachusetts Lowell, University of New Hampshire	Center for High-rate Nanomanufacturing (www.nano.neu.edu)	Sept. 2004
University of California Berkeley	Center for Integrated Nanomechanical Systems (nano.berkeley.edu/coins)	Sept. 2004
Ohio State University	Center for Affordable Nanoengineering of Polymer Biomedical Devices (www.nsec.ohio-state.edu)	Sept. 2004
University of Pennsylvania	Center for Molecular Function at the Nano/Bio Interface (www.nanotech.upenn.edu)	Sept. 2004
University of Wisconsin Madison	Templated Synthesis and Assembly at the Nanoscale (www.nsec.wisc.edu)	Sept. 2004
Stanford University	Center for Probing the Nanoscale	Sept. 2004

Nanoscale Interdisciplinary Research Teams (NIRTs)

NIRT awards are typically for four years and must involve a minimum of three principal investigators. These awards typically fund research for which promising preliminary results have already been obtained and where joining researchers having different expertise will boost the rate of progress and degree of impact. For industry researchers looking at cutting-edge projects with shorter time horizons, NIRTs are a good resource. Typically about 50 to 70 awards are funded each year.

Nanoscale Exploratory Research (NER)

NER awards are typically for one year only. As suggested by the name, these grants are meant for high-risk, high-impact projects at a very early research stage. About 50 to 70 are funded each year.

In addition, NSF has funded two user networks. These networks are meant to provide access for both industrial and academic researchers to instrumentation, tools, and computational resources.

National Nanotechnology Infrastructure Network (NNIN)

The NNIN (www.nnin.org), funded at a level of $14 million per year for five years and renewable for another five years, was completed in 2004 after conclusion of the National Nanofabrication Users Network (NNUN). Led by Cornell, the initial network includes Georgia Institute of Technology, Harvard University, Howard University, North Carolina State University, Pennsylvania State University, Stanford University, the University of California at Santa Barbara, the University of Michigan, the University of Minnesota, the University of New Mexico, the University of Texas at Austin, and the University of Washington, with the potential of adding new nodes. The mission of the NNIN is to provide access to nanotechnology-relevant infrastructure for synthesis, characterization, fabrication, and integration.

Network for Computational Nanotechnology (NCN)

Initially funded in 2002 and led by Purdue, the NCN (ncn.purdue.edu and www.nanohub.org) has a number of participants, including Northwestern University, University of Illinois, University of Florida, Morgan State University, University of Texas at El Paso, and Stanford University. The mission of the NCN is to create and provide access to new algorithms, modeling approaches, software, and computational tools.

Education and Workforce Development

In the 2003 NSF solicitation, a new component in Nanotechnology Undergraduate Education (NUE) was added, followed by a separate solicitation on Nanoscale Science and Engineering Education (NSEE) in 2004. In addition to the funding provided by the NUE, the NSEE included funding for Centers for Learning and Teaching (NCLT), Informal Science Education (NISE), and Instructional Materials Development (NIMD). These efforts are focused primarily on expanding nanotechnology knowledge and generating excitement in K–12 and undergraduate students.

NNI AND NSF RESEARCH FOCUS AREAS

One of the strengths of nanotechnology is its potential to impact a wide range of technology areas, but it is often easier to understand and foresee probable areas of scientific advancement by recognizing the organization of the funding initiatives. For example, the following represent various sets of R&D focus areas that have been identified over the past five years.

The original National Nanotechnology Initiative plan identified nine "grand challenges."[1] Significant emphasis was placed on connections to the interests of the participating federal agencies, to establish major, long-term objectives in nanotechnology in these areas:

1. Nanostructured Materials by Design
2. Nano-Electronics, Optoelectronics, and Magnetics
3. Advanced Healthcare, Therapeutics, and Diagnostics
4. Nanoscale Processes for Environmental Improvement
5. Efficient Energy Conversion and Storage
6. Microspacecraft exploration and Industrialization
7. Bio-nanosensors for Communicable Disease and Biological Threat Detection
8. Economic and Safe Transportation
9. National Security

Later modifications to the list included two additional grand challenge areas: Manufacturing at the Nanoscale and Nanoscale Instrumentation and Metrology. In addition, grand challenges 7–9 were combined into one category: Chemical-Biological-Radiological-Explosive Detection, and Protection. The most recent strategic plan, "The National Nanotechnology Initiative: Strategic Plan"

(December 2004), identified a slightly different organization of seven program component areas for investment:

1. Fundamental nanoscale phenomena and processes
2. Nanomaterials
3. Nanoscale devices and systems
4. Instrumentation research, metrology, and standards for nanotechnology
5. Nanomanufacturing
6. Major research facilities and instrument acquisition
7. Societal dimensions

These were based on the stated NNI goals of (1) maintaining world-class R&D; (2) facilitating transfer of new technologies into products for economic growth, jobs, and other public benefits; (3) developing educational resources, a skilled workforce, and the supporting infrastructure and tools; and (4) supporting responsible development of nanotechnology.[2]

The effect of these identified focus areas can be seen in the research themes of the NSF Nanoscale Science and Engineering (NSE) solicitations, the most recent of which identified the following research and education themes:

- Biosystems at the Nanoscale
- Nanoscale Structures, Novel Phenomena, and Quantum Control
- Nanoscale Devices and System Architecture
- Nanoscale Processes in the Environment
- Multi-scale, Multi-phenomena Theory, Modeling and Simulation at the Nanoscale
- Manufacturing Processes at the Nanoscale
- Societal and Educational Implications of Scientific and Technological Advances on the Nanoscale

On a broad scale, the NSF research themes represent the interests of the NSF and its program directors as well as input from the technical community through strategic workshops and submitted proposals. Thus, large groupings of research discoveries are likely to be organized under these themes. Therefore, if a decision maker in, say, a biomedical sensor company wanted to find out what discoveries might be applicable to its products, the decision maker would not only want to look at awards under the Biosystems at the Nanoscale theme area, but also Nanoscale Devices and Manufacturing at the Nanoscale. Many of the research projects may have a particular "test bed" application in mind, but with nanotechnology, as with other research, some of the critical developments are actually applications of technology from one field to a different field.

Following are brief descriptions of the theme areas excerpted from a NSF solicitation.[3]

- **Biosystems at the Nanoscale.** This theme addresses the fundamental understanding of nanobiostructures and processes, nanobiotechnology, biosynthesis and bioprocessing, and techniques for a broad range of applications in biomaterials, biosystem-based electronics, agriculture, energy, and health. Of particular interest are the relationships among chemical composition, single-molecule behavior, and physical shape at the nanoscale and biological function level. Examples include the study of organelles and subcellular complexes such as ribosomes and molecular motors; construction of nanometer-scale probes and devices for research in genomics, proteomics, cell biology, and nanostructured tissues; and synthesis of nanoscale materials based on the principles of biological self-assembly.
- **Nanoscale Structures, Novel Phenomena, and Quantum Control.** Research in this area explores the novel phenomena and material structures that appear at the nanoscale, including fundamental physics and chemistry aspects, development of the experimental tools necessary to characterize and measure nanostructures and phenomena, and development of techniques for synthesis and design. This research is critical to overcoming obstacles to miniaturization as feature sizes in devices reach the nanoscale. Examples of possible benefits include molecular electronics, nanostructured catalysts, advanced drugs, quantum computing, DNA computing, the development of high-capacity computer memory chips, production of two- and three-dimensional nanostructures "by design," nanoscale fluidics, biophotonics, control of surface processes, and lubrication.
- **Nanoscale Devices and System Architecture.** Research in this area includes development of new tools for sensing, assembling, processing, manipulating, manufacturing, and integration along scales; controlling and testing nanostructure devices; design and architecture of concepts; software specialized for nanosystems; and design automation tools for assembling systems of large numbers of heterogeneous nanocomponents. One can envision "smart" systems that sense and gather information and analyze and respond to that information; more powerful computing systems and architectures; and novel separation systems that provide molecular resolution.

- **Nanoscale Processes in the Environment.** Research in this area will focus on probing nanostructures and processes of relevance in the environment from the Earth's core to the upper atmosphere and beyond. Emphasis will be on understanding the distribution, composition, origin, and behavior of nanoscale structures under a wide variety of naturally occurring physical and chemical conditions, including nanoscale interactions at the interface between organic and inorganic solids, between liquids and gases, and between living and nonliving systems. Examples are biomineralization of nanoscale structures, molecular studies of mineral surfaces, study of the transport of ultrafine colloidal particles and aerosols, and study of interplanetary dust particles. Possible benefits include gaining a better understanding of molecular processes in the environment, developing manufacturing processes that reduce pollution, creating new water purification techniques, composing artificial photosynthetic processes for clean energy, developing environmental biotechnology, and understanding the role of surface microbiota in regulating chemical exchanges between mineral surfaces and water or air.
- **Multi-scale, Multi-phenomena Theory, Modeling and Simulation at the Nanoscale.** The emergence of new behaviors and processes in nanostructures, nanodevices, and nanosystems creates an urgent need for theory, modeling, large-scale computer simulation, and new design tools in order for researchers to understand, control, and accelerate development in new nanoscale regimes and systems. Approaches will likely include and integrate techniques such as quantum mechanics and quantum chemistry, multiparticle simulation, molecular simulation, grain- and continuum-based models, stochastic methods, and nanomechanics. Of particular interest is the interplay of coupled, time-dependent, and multiscale phenomena and processes in large atomistic and molecular systems to make connections between structures, properties, and functions. Examples of possible benefits include the realization of functional nanostructures and architectures by design, such as new chemicals, multifunctional materials, bioagents, and electronic devices.
- **Manufacturing Processes at the Nanoscale.** Research in this area will focus on creating nanostructures and assembling them into nanosystems and then into larger-scale structures. This research should address these areas: understanding nanoscale processes; developing novel tools for measurement and manufacturing at the nanoscale; developing novel concepts for high-rate synthesis and processing of nanostructures and nanosystems; and scale-up of nanoscale synthesis and processing methods. Examples are synthesis of nanostructures for various functions,

fabrication methods for devices and nanosystems, design concepts for manufacturing, simulation of the manufacturing methods at the nanoscale, and evaluation of the economic and environmental implications of manufacturing at the nanoscale. Possible benefits include improving understanding of manufacturing processes in the precompetitive environment, generating a new group of nanoscale manufacturing methods, increasing the performance and scale-up of promising techniques, and establishing the physical and human infrastructure for measurements and manufacturing capabilities.

- **Societal and Educational Implications of Scientific and Technological Advances on the Nanoscale.** Innovations in science and technology require societal support and also influence social structures and processes, sometimes in unexpected ways. We need to examine the ethical and other social implications of these societal interactions to understand their scope and influence and to anticipate and respond effectively to them. Support for nanoscience and nanotechnology is likely to enhance understanding of fundamental natural processes, from living systems to astronomy, and to change the production and use of many goods and services. Studies of the varied social interactions that involve these new scientific and technological endeavors can improve our understanding of, for example, the economic implications of innovation; barriers to adoption of nanotechnology in commerce, health care, or environmental protection; educational and workforce needs; ethical issues in the selection of research priorities and applications and in the potential to enhance human intelligence and develop artificial intelligence; society's reaction to both newly created nanoparticles and nanoparticles that newly developed techniques permit us to recognize, detect, characterize, and relate to health and environmental issues; implications of the converging interests of different fields of science and engineering in the nanoscale; risk perception, communication, and management; and public participation and involvement in scientific and technological development and use. This theme aims to develop a long-term vision for addressing societal, ethical, environmental, and educational concerns.

Along with these themes, a Memorandum of Agreement with SRC (Semiconductor Research Corporation, which sponsors university research on behalf of its member companies) led to the addition of a seventh theme area: Silicon Nanoelectronics and Beyond. This research, which represents an example of a joint collaboration between NSF and industry, encourages proposals of interest to a particular industry sector through its inclusion as a theme area in the NSF solicitation, while providing additional funding opportunities from SRC funds. Similar joint solicitations have been arranged between NSF and other federal agencies, such as DoD, DOE, and EPA. Many of these funding agencies have searchable databases that identify funded research in specified theme areas.

FUTURE DIRECTIONS

The nature of federal funding of academic research can be envisioned as tossing pebbles into a lake. Major research program solicitations tend to run for about five years, providing only the initial impact of the pebble on the water, with the expectation that the spreading rings will last much longer and spread much farther. The next pebble can fall in a completely different part of the lake, or it can fall in an overlapping area of the spreading rings, leading to interaction between rings in a region. Because of its impact, nanotechnology has already been given more pebbles to work with; nanotechnology support was extended beyond five years and further supported by Congress in December 2003 with the enactment of the 21st Century Nanotechnology Research and Development Act.[4]

Although the funding for nanotechnology will logically continue to shift from its current form, industrial interests are more likely to focus on the subsequent outcomes of earlier funded research. The hope, with concurrent funding in Manufacturing at the Nanoscale, is that the typical period that elapses between initial discovery and widespread commercialization—some 20 to 30 years—can be shortened for nanotechnology.[5]

In the next few years, as commercial products and processes become more viable and more visible to the general public, the emphasis on environmental and health implications will grow. Already, federal agencies such as NIH, National Institute of Environmental Health Studies (NIEHS), NIOSH, and the EPA are establishing new multimillion-dollar initiatives in exposure, transport, detection, and toxicity of nanoparticles and other nanomaterials.[6]

Many of the initial discoveries were rooted in the extension of the quantum and molecular scales of physics and chemistry and the reduction of the microscales of microelectronics, but the most exciting future discoveries lie in the intersection and integration of several technologies and scales. Introduced even in the earliest NNI literature, this convergence of nanotechnology with biotechnology, information technology, and cognitive sciences is anticipated to lead to significant advances in human health, human performance, human productivity, and quality of life.

Such grand challenges are important because they ensure coordination among researchers from a wide range of backgrounds. Perhaps one of the most far-reaching benefits of the NNI has been its effectiveness in encouraging scientists, engineers, and social scientists to learn how to communicate with each other.

RESEARCH FOCUS AREAS OF OTHER NNI AGENCIES

Many other agencies involved in the NNI are active in supporting academic research in nanoscale science and technology.

The Department of Defense has funded research in nanoscale materials and applications that would meet the operation needs of the U.S. Armed Forces. These include developing nanostructures for revolutionary catalysts and sensors as well as developing nanoscale applications in areas such as solar cells, solid-state power generation, and decontamination from chemical, biological, or radiological attacks. DoD also funds a substantial share of the NNI investment in nanoelectronics, nanophotonics, and nanomagnetics research, in particular research on spin electronics and quantum information science. The U.S. Army's Institute for Soldier Nanotechnologies at MIT provides a vital conduit for the transition of academic innovations in nanoscale science and engineering into applications for the benefit of warfighters in the field. DoD agencies have also funded a number of Multi-University Research Initiative (MURI) topics in nanotechnology-related areas.

The Department of Energy has funded a wide variety of research efforts in nanotechnology. These include fundamental research into the unique properties and phenomena that matter exhibits at the nanoscale. DOE research interests include catalysis by nanoscale materials, the use of interfaces to manipulate energy carriers, nanoscale assembly and architecture, and scalable methods for synthesis of nanomaterials.[7] DOE has also funded extensive research into modeling and simulation of nanoscale materials and processes, as well as advanced computing to enable large-scale modeling and simulation.

DOE's network of five NSRCs will play a key role in providing access to large-scale research facilities and equipment needed by academic nanotechnology researchers. Small universities in particular do not always have the resources to purchase the latest nanotechnology research instruments, and even large institutions may find it prohibitive to buy specialized light sources and other large-scale instruments that are needed for some experiments. These resources will be available at the NSRCs, which are user facilities available to the entire R&D community, with time and staff support allocated on the basis of merit-reviewed proposals. The centers are under construction as of this writing, and are co-located with complementary light sources and other facilities at Oak Ridge National Laboratory, Argonne National Laboratory, Brookhaven National Laboratory, Lawrence Berkeley National Laboratory, and a combined center at Sandia National Laboratory and Los Alamos National Laboratory. Time, materials, and technical assistance are provided free of charge for researchers who are willing to publish the results of their work in the open literature, and on a cost-recovery basis when used for proprietary purposes. These DOE centers complement NSF's National Nanotechnology Infrastructure Network and the Network for Computational Nanotechnology, NNIN's network of user facilities devoted to computational modeling of nanomaterials.

The National Institutes of Health has many Institutes that are funding research into nanotechnology-based disease detection, prevention, and therapeutics, as well as tools that will allow greater understanding of the biological processes that influence our health. Nanotechnology research can develop unique capabilities to measure and manipulate on the size scale of individual cells. Efforts have already been funded to develop new tools that gather information on macromolecular systems in living cells or organisms. Other funding is targeted to detection and treatment of specific diseases—for example, research into targeted imaging contrast agents for detecting very small numbers of cancer cells at the earliest stages of the disease, and even nanoparticles that can selectively deliver chemotherapy drugs only to tumors without damaging normal tissues. The National Institutes of Health, like most federal agencies, funds considerable other basic nanoscale research that it does not formally classify as nanotechnology.

The Environmental Protection Agency is funding research in nanotechnology applications that protect the environment by sensing and removing pollutants as well as research in more environmentally friendly manufacturing technologies. EPA also is supporting extensive research studies on how nanoscale materials might impact human health and the environment. In particular, in 2004 EPA took the lead in organizing (with NSF and the National Institute for Occupational Safety and Health, or NIOSH) an interagency program for environmental and human health effects of manufactured nanomaterials. This project is likely to be expanded in future years, possibly to include funding agencies in other countries.

NASA sponsors a wide variety of research on the fusion of nanotechnology, biotechnology, and information technology to study and mimic the organizational structures found in nature. Other research funded by NASA supports space exploration in areas such as sensors for extraterrestrial materials and applications that monitor human health. NASA researchers have collaborated with universities in a number of nanotechnology-related areas, including some leading-edge research in carbon nanotube fabrication and applications. In 2003 and 2004 NASA funded four University Research, Engineering and Technology (URETI) centers devoted to promoting interdisciplinary academic research in a variety of nanotechnology-related topics. These are located at UCLA, Texas A&M University, Princeton University, and Purdue University. The initial awards were for five years, with options to extend NASA's support for an additional five years.

SUMMARY

The ability to take advantage of the products of academic research lies in understanding the structure of academic research funding. Depending on whether the goal is to obtain a general introduction to a new technical area, to identify technologies that are close to commercialization, or to keep up with revolutionary new ideas, one can narrow the starting point for a search by identifying appropriate centers, groups, or individual investigators through award databases.

In a more active mode, industry can have a guiding and accelerating influence on the progress of academic research, either through participation in workshops and conferences that identify the critical barriers or through true research partnerships with academia, supporting not only the research but also the development of the students who will constitute the future workforce.

REFERENCES

1. "National Nanotechnology Initiative: The Initiative and Its Implementation Plan," NSTC/NSET report, July 2000.
2. The National Nanotechnology Initiative Strategic Plan, December 2004.
3. http://www.nsf.gov/pubs/2004/nsf04043/nsf04043.htm.
4. http://www.nano.gov/html/about/history.html.
5. http://www.nsf.gov/crssprgm/nano/.
6. http://www.nano.gov/html/facts/EHS.htm.
7. http://nano.gov/nni_energy_rpt.pdf.

Chapter 8

Understanding University Technology Transfer for Nanotechnology

Larry Gilbert and Michael Krieger

At first blush a straightforward process, commercializing the results of university research depends on many factors coming together "just right," and in a hospitable environment. Because the variability of the actors and elements in the process tends to dominate any idealized model of successful technology transfer, this article focuses the unique features of the components rather than the process per se.

Universities are a wellspring of nanotechnology discovery, with the dominant portion supported by the federal government as it puts ever-increasing public funding into nanotechnology grant and contracts programs.[1] In contrast, corporate and other private developments are just that—private and substantially unavailable to entrepreneurs, investors, or others in the public. Combining this dichotomy with the fact that pursuing most nanotechnology ideas requires a level of capital investment and a team of scientific talent rarely assembled by the lone inventor, the research fruits of universities (which typically are mandated by federal funding to be made available for commercialization) present a central opportunity to entrepreneurs and investors for involvement in nanotechnology—to find discoveries to turn into viable commercial products.

The lab-to-market journey—whether involving nanotech or another technology—is generally referred to as *technology transfer* and is initially shepherded by a university's office of technology transfer (OTT) or licensing office.[2] Idealized, that process is often characterized in terms of these major steps:

1. A research discovery or technology advance is made by a professor or other senior researcher (the "invention").

2. The invention is disclosed to the university's technology transfer office.
3. That office files a provisional patent to protect associated intellectual property (IP).
4. The office "markets" the invention to or responds to commercial interest from potential licensees that are willing to bear the cost of patenting the invention, developing the technology to commercial viability, and taking it to market.

Three caveats are in order lest this process seem overly straightforward and easy to execute.

First, the sequence requires that all the actors involved—the people, companies, and institutions—pull together, that is, cooperate to a considerable degree to make the deal, as well as the technical development, commercially viable. Any significant aberration, such as unreasonable licensing terms, stands to break the chain of steps needed to bring the discovery to market, or perhaps even prevent making it to the patent office. Second, discoveries also advance from lab bench to market shelf in other ways, several of which are described later in this chapter.

Finally, and most fundamentally, we put "markets" in quotation marks because it implies and represents the common belief that an OTT's market push creates technology awareness, interest, and ultimately license deals. In our view, this almost universal faith is misplaced: Although "push" marketing is done almost universally by tech transfer offices nationwide, we believe its effectiveness for sophisticated technologies is minimal, if not mythical, when compared with the "pull" of awareness created by a researcher's publications, professional presentations, former students, Web sites, and so on.[3]

This chapter describes key facets, dimensions, and pitfalls of technology transfer that need to be understood by anyone hoping to participate in moving an attractive scientific or technology discovery into the commercial world.

UNIQUE RATHER THAN COMMON ELEMENTS CHARACTERIZE TECH TRANSFER DEALS

As already suggested, our present focus is not the process per se but rather its components. The reason is fundamental yet perhaps too little appreciated: Although each tech transfer commercialization certainly reflects and results from an identifiable process, it is decidedly not manageable as such. That is, the word *process* suggests that one can delineate a set of steps, criteria, measures, inputs, outputs, and so on and then hand them, as if a recipe, to someone of reasonable skill to manage for a variety of situations and technologies.

That would be true if the commonalities existing among the various successes dominated the story. *Nothing could be further from the truth.* Rather, the unique differences in the constituents must be dealt with. Each deal must be handled with its own particular emphasis on the blend of ingredients, from human factors to royalty rates.

Indeed the misconception of emphasizing tech transfer as a process has seriously undercut the potential success of many tech transfer efforts; unfortunately, emphasizing commonalities of process from deal to deal diverts attention from the unique, vital, and critical deal-dependent differences of the components. These include faculty personalities, the nature of the technology, the time and capital required to advance from laboratory discovery to proof of concept, the nature of the market, and other factors.

Ultimately determinative of success are the motivations and relationships of those involved. We thus look at the researchers, their relationships with their institutions and administrators, and the types of arrangements they have with commercial entities. Subsequent sections treat those components and their special elements that affect the success of transferring an invention and technology.

WHY DOES A UNIVERSITY TRANSFER TECHNOLOGY?

This question is underscored if one looks at the statistics: Few university OTTs generate enough income to justify themselves when measured strictly by revenue. But many other reasons exist:

- **Federal Mandate.** U.S. government funding may require the university to at least make the resulting discoveries available to those who might commercialize them.
- **Equity.** A share in an early-stage company that succeeds can eventually be very valuable.
- **Faculty recruitment.** A significant number of faculty members want to see their results commercialized, for reasons that range from the greater social good, to the opportunity to interact with industry, to the hope of enhanced income from sharing in royalties resulting from the invention's commercialization.
- **Goodwill.** Faculty and alumni who become business successes, and others in the business world, are more likely to become donors if they recall and see an environment that encourages technology entrepreneurship and business development.[4]

HOW IS TECHNOLOGY TRANSFERRED?

Technology can pass from a university into the commercial world in many ways; five are discussed in more detail later in the section "Types of Business Relationships." Typically, the discovery or result of interest centers on a patentable invention by a professor for which the university will file a provisional if not full utility patent. The simple process model sketched earlier then posits that a commercial entity learns of the invention, desires to exploit the technology, and thus executes a licensing arrangement with the university. The licensee may be an established company or a start-up formed by entrepreneurial researchers from the university. Depending on the type of licensee, the agreement may provide for some up-front payment or equity in the licensee in addition to royalties on sales when the technology enters the market.

In reality this simple picture is incomplete or masks many potentially distorting elements. Most fundamentally, successful development of a technology usually requires—in addition to rights in the relevant patents—the know-how of the discoverer and his or her graduate students and other lab personnel. This in turn means that the licensee's commercial venture likely requires the professor's cooperation, typically in the form of a consulting arrangement or employment of (former) students. It is essential to appreciate that absent such know-how a patent may well be worthless.

With this in mind, we now look at the sources of technology, the effects of academic culture, and the types of business relationships that provide vehicles for technology transfer.

Sources of Technology

Implicit in the preceding section is a licensable or otherwise transferable technology, that is, one that has come to the attention of the university and is being protected by the patent system.[5] Indeed, by virtue of faculty, grad student, and staff employment agreements, as well as the Bayh-Dole Act requirements for federal funding, ownership in any invention—and typically in any discovery—vests in the university (unless it disclaims the invention).[6]

How does the university administration learn of the invention? It finds out in the same way corporate administrations do: by the inventor submitting an invention disclosure form. Of course, that depends upon whether the inventors submit the disclosures.

Indeed, to the extent that faculty fail to disclose technology so that patent protection can be invoked, the technology may become part of the public domain and fair game for all, especially if it is disclosed at a professional conference. Although this may at first seem socially desirable, the lack of a patent-conferred monopoly may hinder or effectively prevent commercial development should the technology require a deep investment to render it viable.

What Are the Incentives for the Discloser?

In enlightened companies there are educational programs to help researchers realize that what their deep familiarity with the technology makes "obvious" may not be and could well be patentable. Moreover, many companies conspicuously display their patents and reward their inventors with payments upon invention disclosure, patent filing, and patent issue.

In contrast, few universities have effective IP education programs, and virtually none has an immediate cash award. But in contrast to private industry, university inventors typically receive a substantial share of the royalty stream, commonly 25–50 percent (an arrangement very rare in the private sector).

Faculty Trust

Although such a high royalty percentage might seem to be a strong incentive for invention disclosure, faculty generally recognize that the tech transfer policies and practices of their institutions are a gateway to achieving royalties. Thus their confidence—or lack of it—in their tech transfer offices is a key determinant of their willingness to file the disclosure statements that trigger the OTT's involvement and IP protection. After all, why bother if nothing will come of it?

More broadly, lack of trust may arise for various not entirely independent reasons:

- Distrust of institutional administrators generally
- The belief that tech transfer officers are looking out for the university and might at some level be disingenuous with the faculty member
- Doubt about the OTT's ability to market (find licensees for) the invention (for researchers who believe this to be the OTT's role)
- Doubt that the OTT has the business sophistication and nimbleness to make contract arrangements that investors or industry will accept
- An OTT's reputation (whether or not justified) for being difficult, unresponsive, or unsuccessful

It should be recognized that a certain amount of tech transfer will take place in every major research institution, even those saddled with these liabilities. Some technologies speak for themselves, some researchers have enormous entrepreneurial and business skills, and some individual technology transfer officers have the interpersonal skills to overcome a general distrust of their offices or their histories.

But in the institutions that have had truly successful programs, the researchers trust the OTT (and conversely the office trusts them). Not only does this encourage formal disclosure of inventions, but also it means, among other things, that the OTT is likely to be privy to research developments early on, is welcome in the lab, meets the graduate students who may be the ones to carry the start-up ball or to go to work for a licensee, and so on. Furthermore, it means that when the OTT wants to show or explain the technology to a potential licensee or partner, the researcher is likely to cooperate and spend the time it may take.

Academic Culture Issues

A potential licensee, especially a corporate entity, may have to make considerable adjustment in dealing with a university compared to the familiar styles of commercial suppliers and partners. This can manifest itself, for example, in problems with maintaining the secrecy it may usually take for granted and in the slow pace of negotiation. In the following sections we delineate key factors affecting the perspectives of university personnel.

Faculty Publication

"Publish or perish" remains alive and well. A professor's publication record is central to salary and position within the university. Similarly, any researcher's bibliography is a critical qualification for external funding. In parallel, reputation among peers—arising from activities such as research collaboration and conference presentations—is requisite for the professor to get the favorable letters of support solicited by outside funding sources (such as federal agencies) and by campus promotion committees. All this puts a premium on early dissemination of research. Closely allied is the spirit of academic freedom—the belief that on campus almost any topic of discussion is acceptable and available to be discussed and shared freely—and its corollary: Restraints on sharing information are unacceptable.

What does this culture imply? First, it implies that commercially valuable ideas may be released into the public domain, that potential patents may be barred by early disclosure, or that an existing licensable patent (or application) may be subject to invalidation. Second, if an acquirer of technology is considering a partnering arrangement there will need to be a mutual assessment of what level of research secrecy is really needed and how it fits the personnel and institution involved; for example, it must be determined that no insurmountable conflict exists with the views of academic freedom held by the campus personnel, with university policy, or with obligations under government funding.

Practical needs for publication and presentation must be taken into account when a commercial entity negotiates consulting services, whether they are made directly with a faculty member or as part of licensing a patent. To this end, for example, confidentiality provisions may need to be customized in lieu of using nondisclosure agreement (NDA) forms. More generally, to the extent that key faculty are necessary to exploit a technology, personality issues come into play; some university faculty are resistant to being told what to do; others are quite naive about business practices, and still others are most cooperative. The main issue is to be prepared so that, for example, cooperative but less worldly faculty can be coached to appreciate corporate business practices, to understand that some of their work for the company cannot be immediately shared at conferences or in publications or that turning research into a marketable product takes many times the cost of the underlying discovery and will take time.

Technology Transfer Offices and University Administration

To a significant degree, the nature of a university or other research center as a large organization is antithetical to developing a successful technology transfer program. In large measure the conflict is one of having a short- versus a long-term view in establishing goals and evaluating results. Although many of the reasons for technology transfer are intangible benefits, an OTT nonetheless costs money, so licensing revenue becomes the most natural measure of its success.

To a start-up, licensing means little or no initial royalty payment and a long window until sales yield royalties. And even if an established corporation is the licensee, it is likely to require some years before profitable, royalty-generating sales occur. Meanwhile the costs of filing and maintaining a patent portfolio as well as the OTT staff and overhead costs appear every year in the university budget. This may lead to tension between the OTT staff and university administration as to whether or not to patent an invention; for example, the admin may want to patent only the "viable" technologies (somehow expecting technology transfer personnel to bat 1000 at picking the winners). Yet picking a commercial success at such an early stage is simply not possible. To underscore the point, venture capitalists—who invest in technology at a much later stage—themselves end up funding far more losers than successes.

On the other side of the coin, many researchers will believe their discoveries are unquestionably worth patenting, and then they will become disenchanted with the OTT should it fail to recommend patenting their discoveries.

Another organizational factor running counter to tech transfer effectiveness is that OTT personnel may feel pressure to perform in a shorter time frame than is realistic; in other words, annual reviews do not comport with the three to five years it may take to know whether the deals done by the OTT member were good. This can cause turnover of tech transfer staff and the attendant failure to build the faculty relationships and develop the trust that is requisite to motivating invention disclosures and post-disclosure cooperation.

Types of Business Relationships

Although licensing a patentable technology to a commercial entity is the most visible mode of transferring technology, it is only one of a variety of arrangements that are commonly done. Valuable IP is typically acquired through one of the following forms of relationship with research institutions.

Licensing

In this first method, the university grants certain rights in intellectual property to a third party, the licensee. Licensees are typically one of two types: an existing company interested in some technology the university holds, or a new start-up company. Licenses can be exclusive or nonexclusive and usually involve patents, although some occasionally include software.[7]

The university will try to limit the field of use for the license, and the company will often want to negotiate a more expansive field because it may be unclear at the time the license is granted which fields or applications best suit commercial application of the invention. The university will generally require, in the case of exclusive licenses, evidence of diligence in exploiting the technology. Generally the lack of such diligence will cause the license to become nonexclusive or the field of use to be reduced.

Even when granting an exclusive license, the university may retain certain rights, including the right to use its invention for research and development. The federal government also retains rights under the Bayh-Dole Act, and there may also be third-party rights.

Failure to investigate and to keep these rights in mind can lead to later disappointments.

Faculty Consulting

As the leaders—and frequently the definers—of cutting-edge scientific breakthroughs and technology advances, university faculty can be uniquely valuable consultants to a corporate R&D effort.

For the most part, patented technology is best developed and exploited with the help of its inventor, because there is usually extensive, valuable know-how that is not necessarily disclosed in the formal patent. To this end, part of a licensing arrangement will involve participation by the faculty members whose research led to the patent.

As with any other outside contractor to the company, faculty consulting may also be arranged independently, in which case there may be no specific university involvement. In this case, particular care must be taken to set out the parties' expectations and ground rules for the scope of the engagement. This is because faculty members almost surely have obligations to the university regarding ownership of IP related to their funded research. If the consulting yields, for example, patentable inventions closely aligned with the university-based efforts, disputes over ownership could arise, especially were the consulting to lead to a corporate product with visible financial success.

Strategic Partnering with University Spin-Offs

Promising, innovative university technologies are often best developed under the mantle of a separate company. The transformation from lab bench discovery to economically viable technology may require resources not found in or not appropriate to the university setting; conversely, the intense and autonomous work needed to prove the technology may not fit under the corporate wing. Such a start-up may be initiated by a faculty member and graduate students, or it may be explicitly contemplated by a licensing arrangement. In either case, a technology license and corporate investment in the company will likely go hand in hand. Often, the relationship will have been initiated by corporate R&D personnel, who learn of the technology from conferences and publications. Other times university "show and tell" events give rise to the relationship.

Special Funding of Faculty Research

Once they are aware of faculty working in areas of corporate interest, some companies have found it useful to cultivate a link with the faculty member through research grants. Typically such grants do not have a specific mission but rather simply contemplate work within a specific area. In significant measure, the main function of such grants is to establish a channel of communication with the faculty member so that the company has an early awareness of the research or ready access to consulting time of the professor and employment of advanced and knowledgeable students.

Major and Ongoing Research Partnering

Continuing relationships between a corporation and a university may take many forms. Specific departments and schools often have industrial affiliate programs in which a company makes an annual contribution and, in turn, gains advanced access to university research through special programs, meetings, and the like. At the other end of the spectrum, a company may fund a major research facility on an ongoing basis in exchange for a right of first refusal and certain exclusive rights to the fruits of discovery from the funded research.

Risks

Perhaps the biggest risk in acquiring technology from a university is to be sure that the ownership of the IP is clear. Although this is not as likely to be a problem in pure licensing arrangements, it can become murky in consulting and partnering unless one is alert to potential pitfalls. Indeed litigation—although relatively uncommon—has been rising in the tech transfer context, no doubt because of the increased frequency of technology being transferred without a corresponding gain in sophistication about tech transfer and intellectual property. Considered here are the most common litigation risks and what a business can do to prevent getting embroiled.

University Licensing Litigation Risks

Inadequate attention to detail when licensing university technology can result in a host of problems. First, a company should require that the university represent and warrant that it owns the technology that it is transferring. There have been cases in which non-university entities have claimed they developed the same technology the university has licensed. Second, make sure that the university is clear about the scope of the technology's application or market. Litigation has arisen over the proper scope of the university's license to an outside company. Third, inquire whether the university has offered other licenses on the technology, to guard against errors in the university's licensing program. On occasion, universities have been known to unwittingly offer overlapping "exclusive" licenses. Similarly, a "favored-nation" clause in a non-exclusive agreement with an initial licensee may be overlooked in deals with later licenses.

Professor Consulting Relationships

Hiring a professor who is an expert in a company's research and development field is a popular and effective way to bring cutting-edge academic knowledge to the corporate setting. Unfortunately, there are litigation risks. First, and foremost, the corporation must assure itself that the professor is not taking university-owned technology off campus. This can be a tricky inquiry. The best place to start is to obtain a copy of the university's patent policies. This will give the corporation a checklist of the do's and don'ts for its consultant professor.

Typically, a university's patent policy is part of its professors' binding employment contracts. The corporation should ask professors to represent and warrant that none of the work that they will be doing for the corporation would represent a violation of their obligation under the patent policy. A recommended precaution is to be sure that professors in fact understand the policy by, for example, having corporate counsel review that policy with them.

Second, the corporation should make sure that professors have not entered into any joint venture with any other person (such as an agreement with a former graduate student to develop the same technology with which the professor is now helping the corporation). Again, the best way to protect against this litigation risk is to have professors represent and warrant that the technology that they are consulting about has not been developed in conjunction with (or with the assistance of) any third party.

Finally, a consulting professor can cause problems when the corporation is making a bid for university technology. Although it is true that a university in a technology licensing bidding process may well favor a company that has hired one of its faculty members, the professor (and his or her corporation) must be careful not to corrupt the bidding process. Professors should not use their position at the university to influence the bidding on a university patent license beyond the channels available and known to every bidder. For example, professors should not use the faculty dining room as an opportunity to politic the head of the licensing office for the company to get the technology when other bidders cannot engage in the same sort of lobbying. Such activity could, in some jurisdictions, subject both the professor and the company to legal liability.

Other Litigation Risks

Because universities are increasingly attempting to market their technology, a corporate researcher may be shown technology under circumstances that suggest confidentiality or other obligation, and thus the researcher needs to be careful about how information gleaned from visiting university facilities is used. Overzealous corporate researchers could create significant exposure to liability if they misuse university technology learned about while chatting with university faculty.

In sum, as with all corporate dealings, the basic rules of the road are the same for avoiding litigation in technology transfer licensing: (1) Get it in writing, (2) make sure somebody else's intellectual property rights are not being violated, and (3) make sure that deal terms are clear and state exactly who bears the risk.

FINAL WORDS

At the heart of a successful tech transfer program is the trust between the faculty and the university personnel shepherding the invention toward viability and the commercial world. Without that trust, far less technology will be disclosed by researchers, faculty will be less willing to cooperate with the OTT, and unseen "leakage" of inventions into the private sector will be more common. Although our setting is academic and not corporate, it is appropriate to recall the words attributed to John Patterson, founder of NCR Corporation, who said nearly a hundred years ago that the only assets a company really has are people, ideas, and relationships.

NOTES

1. Because universities are the dominant component of the nation's public research enterprise, we use the terms *university* and *academic* as a collective that also includes noncommercial academic and think-tank research centers and certain nonprofit centers formed by industry consortia, to which most of the observations in this chapter also apply.
2. The term *technology transfer* is sometimes used more broadly to embrace most any licensing of technology between entities, such as a catalytic process licensed by a major chemical corporation to a major refiner. We use it here in the narrower sense described earlier.
3. Caltech's OTT does no traditional technology marketing and yet is widely acknowledged to have among the country's most fertile tech transfer programs, despite the institute's small faculty.
4. Indeed, very large donations exceed amounts that any successful technology ever can be expected to yield directly. For example, Broadcom co-founder Henry Samueli contributed $30 million to the UCLA Engineering School, and venture capitalist Mark Stevens gave $22 million to USC for an Institute of Technology Commercialization. Inventor and businessman Arnold Beckman has contributed in excess of $100 million to Caltech and large sums to other schools as well, and Gordon Moore's commitments to Caltech exceed $500 million.

5. In the commercial world, patents are typically born as trade secrets by virtue of the implicit status of private information held by a company's researchers. In the academic world, one rarely counts on such protection for two intertwined reasons: The spirit of the academy has a big component of sharing knowledge, and so any initial discovery is soon out of the bag—for example, by disclosure at scientific symposia or by exchange between colleagues far and wide; moreover, participation in a lab and its discussions can be very fluid, so not only the number but also the diversity of people involved (students, visitors from other universities, lab techs) makes keeping secrets problematic. By contrast, effective legal enforcement of trade secret protection requires that the owner have treated the subject matter as a trade secret, which means, at least in court, a showing of reasonable security procedures and signed NDAs with all who have had access to the information. Relatively few university research facilities are likely to withstand the scrutiny of such a review.
6. The bulk of research funding for universities comes from the U.S. government and is subject to the 1980 Bayh-Dole Act (35 U.S.C. §§ 200-212). If a university does not want to carry the burden of prosecuting the patents associated with a discovery, it must timely so inform the U.S. government, which may choose to patent the invention.
7. Under the Bayh-Dole Act, where the research was at least partially supported by federal funding, the U.S. government retains significant usage rights even if the license agreement from the university to a third party is for exclusive rights.

Chapter 9

Intellectual Property Policy and Impact

Chinh H. Pham and Charles Berman

Over the past two decades, the ability to manipulate, control, and measure matter at the atomic scale has led to interesting discovery of novel materials and phenomena, discovery that underlies the multidisciplinary areas of research and development known today as nanotechnology. Nanotechnology, simply put, involves the working and manipulating of matter on a scale that is one billionth of a meter in size.[1] The field spans the spectrum of research and development across many disciplines, including physics, engineering, electronics, materials, molecular biology, and chemistry.

Because the market for nanotechnology-related products and services promises to increase significantly, and with more companies trying to stake out territories while grabbing as much of the intellectual property pie as they can, the value and strategic importance of protecting intellectual property have become crucial in the quest to capture valuable opportunities.

According to the ISI Science Citation Index, in 1987 there were only 237 scientific articles published with the prefix *nano-* in the title. By 2002, the number had risen to about 10,000. In addition, between 1999 and 2002, according to the Derwent Patents Index, the number of U.S. patents published for nanotechnology almost doubled. A quick search of the U.S. Patent and Trademark database shows that at the end of 2004, there are about 7,780 issued patents and about 9,060 pending applications containing the prefix *nano-*.

To the extent that many companies are now placing a priority on intellectual property (IP), the building of a strong IP portfolio for strategic and economic purposes will therefore play an important role in whether these companies can create for themselves commercially viable opportunities.

TYPES OF IP PROTECTION

The U.S. Constitution, statutes, and common law provide protection for intellectual property rights through patent, trademark, copyright, and trade secret law. For many early-stage companies, patents and trade secrets will be predominantly used in the protection of their IP assets. However, in many instances, and depending on the circumstance, trademarks and copyrights can also act to enhance the value of the IP assets.

Patents

In the United States, patents (that is, utility patents) offer protection for inventions that are directed to an apparatus, a concept, or a method or process that is novel, useful, and nonobvious. It is a contract between the inventor and the government (the U.S. Patent and Trademark Office, or USPTO, also called the Patent Office), whereby the inventor agrees to disclose the invention to the public, and in return, the USPTO grants exclusivity in the invention to the inventor if the invention is deemed new, useful, and not obvious. This exchange is seen as helpful in developing the technological knowledge base that is available to the public, while promoting the sciences and technologies.

In general, a patent does not grant to the owner the right to practice the invention. Rather, a patent provides negative rights, in that it grants to the patent owner the right to exclude others from making, using, selling, offering for sale, or importing the invention. With the recent change in U.S. patent law, the enforceable term of a patent now expires 20 years from the priority date of the patent application from which the patent issues.

Often, after one obtains patents in any technological field, it is essential to then engineer and tune the patents to create a substantially strong and solid portfolio to maximize the return from the patent assets. In other words, the portfolio should minimize the gaps that competitors can design around. The challenge of creating a strong and solid portfolio is equally applicable in the field of nanotechnology.

Protecting a new nanotechnology-related invention can be a difficult process, especially when the resources are simply not available at the early stages to pursue patent protection. Moreover, because nanotechnology is an emerging field, protecting the invention requires a level of knowledge and technical expertise that may be lacking.

Furthermore, some new ideas simply cannot be patented because they do not rise to the level of being novel or being nonobvious. For nanotechnology-related inventions, there can be substantial overlap with existing technology except that the nanotechnology-related inventions involve subject matter on a substantially smaller scale. Other nanotechnology-related ideas may be eligible only for narrow or partial protection from potential competition and imitation. Even if minimal protection can be obtained, such pursuit can often be expensive and time-consuming, with the ultimate result remaining uncertain.

A decision to seek protection for a new nanotechnology-related product or process should therefore be approached with the careful consideration that one brings to other commercial transactions. Just because a new nanotechnology-related product or process may be eligible for patent protection does not necessarily mean that the cost of obtaining and preserving such protection is justified from a business viewpoint. Patent protection is viewed as one part of the successful marketing equation. Before pursuing patent protection, one should consider a few statutory requirements and strategic points.

Trade Secrets

In addition to patents, a nanotechnology company can also use trade secrets to protect its proprietary information and intellectual property. A trade secret, in general, is confidential business, technical, or other information that gives a company a competitive advantage over its rivals. A trade secret need not be novel and requires no formal filing procedure for protection. Its duration can be indefinite, as long as reasonable efforts are made to keep the information secret. If, however, the trade secret is publicly disclosed or is reverse-engineered by a competitor, legal protection of the trade secret is destroyed.

For a nanotechnology-related invention, because of the nature, science, and scale of the technology at issue, it may be difficult to reverse-engineer the nanoscale technology when compared with traditional technology. To ensure maintenance of the trade secret, a company should pay close attention to its activities externally as well as internally. For example, companies should consider using a nondisclosure agreement (NDA) when exploring a potential partnership, collaboration, or investment opportunity.

In addition, because of the potential for high employee movement in this field, a nanotechnology company should consider instituting a company-wide protocol for maintaining the confidentiality of proprietary information. For example, a nanotechnology company can educate its employees on the importance of maintaining confidentiality, can mark proprietary or appropriate information "Confidential," and can inform departing employees of their obligation to keep such information confidential.

Trademarks

Another strategy for protecting intellectual property rights is through the use of trademarks. Generally, trademark rights protect the name or symbol associated with the company in relation to its commercial products or services. The rights afforded to a trademark prevent others from using similar, confusing names or symbols that may take advantage of the goodwill and quality associated with the protected names and symbols. Trademarks can have unlimited duration, as long as the mark is renewed at the appropriate time and is commercially used continuously.

As more companies begin to position themselves to take advantage of the opportunities in the field of nanotechnology, there has been an increasing trend to register, as a trademark, names that include the word *nano.* A brief check of the trademark database at the U.S. Patent and Trademark Office reveals that, at the end of 2004, there are more than 1,600 such registered and pending marks.

Copyrights

Copyrights protect the original expression of an idea, as opposed to patents, which protect the idea. Protection under copyrights attaches immediately when the idea is fixed in a tangible medium, such as paper, compact disc, or hard drive. Copyrights can offer protection for as long as 75 to 100 years for work made for hire, which is generally the case in the field of nanotechnology.

STATUTORY REQUIREMENTS

In the United States, there are several major hurdles that an inventor must overcome before patent protection can be obtained. Specifically, an inventor must be able to demonstrate that the invention is new, useful, and nonobvious. A patent cannot be obtained unless the invention is new. In addition, the invention must not have been known or used by others in the United States, or patented or described in a publication in the United States or any foreign country, before the applicant made the invention. Moreover, if the invention was patented or described in a publication anywhere in the world or in public use or sale in the United States more than one year prior to the date the application is filed, a patent will be denied.

In connection with this last point, it is critical that a patent application be filed with the U.S. Patent Office prior to the one-year anniversary of the public disclosure, use, or sale of the invention. However, unlike the United

States, most countries do not allow this one-year grace period. "Absolute novelty" is typically the requirement. Thus, any public disclosure, use, or sale of the invention will foreclose the right of the applicant to pursue patent protection in these countries. Public disclosure includes presentations, printed and online publications, public meetings, offers for sale, and discussions with someone not under a confidentiality agreement. Accordingly, because nanotechnology-related inventions have far-reaching benefits beyond the borders of the United States, it is imperative that an application be filed prior to any public disclosure, use, or sale.

To determine whether an invention is *novel* (i.e., new), an application is examined by a patent examiner skilled in the technology of the invention and is compared against the available prior art. The examiner makes a thorough study of the application and all the available public information pertaining to its claims. Currently, much of this examination is conducted electronically, and the examiner has access to databases of the USPTO, European Patent Office, and Japanese Patent Office. The examiner also has access to other non-patent literature; however, in practice this is more limited.

Given the interdisciplinary nature of nanotechnology-related inventions, as well as the potential for these inventions to have application in multiple industries, examiners may have a difficult time performing adequate searches. For example, an invention directed to the manufacture of nanoparticles might involve materials science, chemistry, and physics. Furthermore, if the invention involves the use of the nanoparticles for drug delivery or imaging purposes, additional consideration must be given to the fields of molecular biology and optoelectronics.

Moreover, there is no specific examination group within the U.S. Patent Office dedicated to nanotechnology. Consequently, the search process currently falls to other existing art units, many of which are overburdened with the number of applications to be examined. As a result, it is important that the application be provided with an adequate and thorough search of the prior art to ensure that the invention at issue is novel.

The invention must also be nonobvious. It is common for the U.S. Patent Office to reject a patent application as being obvious. In certain instances, it may well be that the invention is something that has never before existed. However, if the Patent Office determines that a hypothetical person having access to the available information concerning that particular field of technology would have "known" how to make the invention (or modify a previous invention to make it), it may reject the patent as being obvious.

The claimed invention must also be useful. This utility requirement is seldom a hurdle encountered by the established disciplines. However, it can be an issue for nanotechnology-related inventions. By way of example, as a nanotechnology start-up tries to fortify the landscape around its core technology, it may file additional patents that recite multiple commercial uses for the invention in a variety of industries without clearly knowing whether such uses may work. In doing so, the company may encounter a risk that the invention lacks utility and the patent application may be denied.

Inventions also must be fully described in an enabling manner. The requirements for an enabling disclosure can provide a unique challenge and must be carefully addressed. The nanotechnology-related invention being disclosed might involve not yet clearly understood phenomena or might be addressed by technology that is being disclosed for the first time in the application. Accordingly, for nanotechnology-related patent applications it is imperative to use an interdisciplinary team approach. A team of experts will be much better able to address and provide solutions across various nanotechnology platforms.

Having a team of experts, all of whom understand the various aspects of the invention, can lead to an efficiently generated work product without any compromise in quality.

The USPTO evaluates its patent examiners primarily according to the number of applications they process. Examiners therefore have no incentive to spend very much time on any one application; it is to their benefit to process applications as quickly as possible.

The examiner typically does not take the time to read and study an application at length. An examiner usually does a quick search to discover all the relevant patents and other public information pertaining to the claims in an application. The examiner sends the applicant an "Office Action" explaining the examination. Invariably this includes a rejection of at least some of the claims as being obvious in view of the material the examiner located when conducting the review. The applicant then responds by explaining and distinguishing each cited reference, thereby demonstrating to the examiner that a patent should be issued. This routine saves the examiner time because the applicant has spent his or her own time figuring out why the references cited by the examiner are not particularly relevant. Claims that were originally rejected by an examiner are frequently later allowed after the examiner has had the opportunity to read the applicant's response distinguishing the invention from the cited references.

The applicant is also obliged essentially to assist the examiner in the examination. That is, the applicant needs to advise the examiner of prior art of which the applicant is aware. This serves at least two purposes. First, the applicant obtains a patent knowing that because the examiner has considered the prior art there will be a presumption of validity of the patent at a later time. This means that the patent can be enforced or licensed with more confidence. A second reason for advising the Patent Office of the prior art is the requirement that if an applicant intentionally withholds prior art from the examiner during examination of the patent, the patent can later be unenforceable.

Enforcing a Patent

By bringing a court action against an infringer—namely, a person who intentionally or unintentionally copies the invention—the patent owner may seek both an injunction against the infringer and the recovery of monetary damages. The scope of the injunction, as well as damages, will be determined on a case-by-case basis and should be adequate to compensate the patent owner for the acts of infringement. In no event should the damages be less than a reasonable royalty for the use made of the invention by the infringer.

An alternative is to offer the infringer a license to make the patented invention. Such a license may include an initial base payment, as well as a royalty for each unit of the patented invention that the licensee would make, use, or sell.

The USPTO is responsible for examining and issuing patents to eligible inventors. For all practical purposes the role of the USPTO ceases upon issuance of the patent. The USPTO does not monitor commercial transactions for the presence of potential infringement, nor does it enforce patent rights against potential infringers once their presence is made known. It is the duty of the owners of the patent to protect their patent rights at their own expense. Moreover, the USPTO does not guarantee the validity of a patent. At any time during its lifetime, a patent may be found by a court to be invalid and hence unenforceable.

The financial cost of enforcing a patent against a potential infringer is highly dependent upon the complexity of the case, but legal expenses alone can easily reach several hundred thousand dollars and often reach millions of dollars. A few attorneys are willing to litigate such cases for a fee contingent upon winning the case. It may be possible to recoup some of the legal costs should the patent holder win, but this prospect is never certain in advance of the court action.

The patent owner may gain economic benefit from the patent by selling it, exclusively or nonexclusively licensing it, or by commercially exploiting the patented invention. For an independent inventor to attempt either the sale or licensing of the patent, the inventor must first identify the potential buyers or licensees. Once a potential buyer is located, the inventor must then place a value on the patent. If the product covered by the patent has not yet been commercially produced, it may be quite difficult to determine accurately the value of the patent. Therefore, the patent owner may wish to license the patent and obtain a royalty on the future commercial exploitation of the patent. The specific terms of each individual license would have to be negotiated with the licensee.

Provisional Application Versus Regular Application

For many nanotechnology start-ups, the availability of capital resources is often a critical issue in their pursuit of patent applications. In the course of managing this limited capital resource, there may be a constant battle between the need to use the capital to hit predetermined milestones and the need to spend it to strategically build and protect the intellectual property assets.

Because it can be costly to pursue patent protection through the filing of a regular utility application, many nanotechnology start-ups have opted to pursue an alternative strategy of filing a less costly provisional patent application as a way to initially protect their intellectual property.

Unlike a regular utility application, a provisional application does not have any formal requirements, and thus it can be quickly prepared and filed relatively inexpensively. Once filed with the Patent Office, however, the provisional application does not get examined; instead, it provides an applicant a way to establish an early U.S. filing date. The provisional application will also preserve the right to foreign filings.

Often, the provisional application is a technical disclosure from the inventor. Thus, care needs to be taken when preparing the provisional application. In particular, the more details put into the provisional application, the better, because the benefit of the provisional filing will extend only to subject matter disclosed in the provisional application. This can be particularly important for a nanotechnology-related invention, because the multidisciplinary nature of the invention may require an additional focus on disclosures that may not otherwise have been initially anticipated. A provisional application, by its nature, is often not as robust as a regular utility application. Thus, the provisional filing procedure should be followed only in cases of emergency (for example, when a deadline is fast approaching or public disclosure is imminent), or when there is a financial restriction. Otherwise, a complete and well-prepared regular utility application should be filed.

The life of a provisional application is one year from the date of filing and cannot be extended. During this period, the invention can be referred to as "patent pending." As indicated earlier, it is not examined. Still, the "pending" status can be advantageously used to deter potential competition. In addition, when a company is seeking an investment or a strategic partnership, the "pending" status can assure the potential investor or partner who is about to take a step toward a risky investment that patent protection has been pursued.

The U.S. patent system is based on a "first to invent" standard, whereas most other countries award patents based on "first to file." This is a unique characteristic of U.S. patent practice and does not exist in other jurisdictions. The U.S. patent system is concerned with granting the patent to the first inventor and not the applicant that is the first to file. There is therefore periodically an evaluation of which of two or more applicants first conceived and diligently reduced an invention to practice (that is, interference process). This evaluation is extremely complicated, costly, and time-consuming. Nonetheless, interferences may arise in highly competitive emerging technologies.

IP Challenges and Impact: Restriction on Innovation

The pace of technology adoption in nanotechnology will follow patterns similar to that of other fields. If the USPTO were to allow broad patent claims it would restrict innovation by allowing a few major players to control a large share of the field and crowd out small, innovative players. Conversely, patent claims that are not granted or that are too narrow will hinder progress by failing to protect innovation and will also hinder investment in the new innovations. The first scenario is more likely in the nanotechnology field, and in some areas a restriction on innovation may soon be created. This can be especially true for inventions involving carbon nanotubes. There are currently an abundance of issued patents, many of which may be overlapping in scope.

The patent system can provide insight into the landscape of innovation. For example, the number of patents that have been issued or the number of applications that have been filed in a particular area of nanotechnology can provide insight into a particular trend or focus, as well as insight into potential areas that may be commercially exploited by owners of those patents. The number of patents or applications, alternatively, can help determine the risks associated with entering such a market in light of the particular patent landscape. Ownership information of patents can also help to determine whether there will be risks within a particular area or application based on the number of potential competitors and the relative positions of intellectual property.

Citation of patents may also be a useful indicator. When a patent has been repeatedly cited by others, such information can determine the licensing potential of that patent. Accordingly, if a company is looking to establish a dominant position in a particular area of nanotechnology, it may look to enhance its portfolio by acquiring those patents that have been repeatedly cited by others.

Alternatively, repeated citations of certain patents by others can be indicative of potential roadblocks to be overcome when a company is looking to expand its technology outward. In such a scenario, the company may want to determine whether there is an opportunity to create a "picket fence" around the cited patents. In other words, the company may want to file its own patent applications around the cited patents; these new applications would be directed to potentially valuable areas that the cited patents may have missed, so as to create a potential cross-licensing relationship with the owners of the cited patents.

Patent claim scope can also be used as an indicator of available opportunities. To the extent that the patent provides broad scope of protection, the number of competitors entering that market can be minimized.

Litigation

Nanotechnology companies will need to address patent litigation issues from the perspective of enforcing their own patents against a competing product or from the perspective of answering a claim that their product has infringed a competing patent. There has been little or no activity to date in nanotechnology-related patent litigation. It is expected that within the next few years there may be a significant increase in the amount of patent litigation as more nanotechnology-related products become commercialized and nanotechnology companies begin to assert their patents.

Note that in contrast to the rise of other new technologies, such as biotechnology, it is not expected that a new body of law will be developed specifically for nanotechnology. Rather, it is expected that certain facets of existing law will ultimately develop to take into account the special nature of nanotechnology-related inventions.

Funded Research

Although plentiful funds have been awarded by the federal government through the various governmental agencies that are part of the National Nanotechnology Initiative, a company should consider the impact of pursuing government grants to further its nanotechnology-related research. Under the Bayh-Dole Act, universities and small business entities can retain intellectual property ownership rights in federally funded research. The government retains a royalty-free license to any patented technology that is federally funded, although this has very rarely (if ever) been used. Any transfer or acquisition of these patent rights requires compliance with certain formalities and can impact or interfere with the growth of a company or its exit strategy.

CONCLUSION

Implementation of at least some of the processes suggested in this chapter can assist in improving business and increasing the value of the intellectual property portfolio. Different industries, including those based on nanotechnology, will have short-, mid-, and long-term technology and business objectives. Accordingly, the patenting objectives for different aspects of nanotechnology can differ for different industry uses. Every business can extract extra revenue from its business knowledge base by strengthening and building on its patentable assets. It is difficult to act on all of these criteria to define a patenting policy, but implementation of at least some of these processes can assist in improving business and increasing the value of your patents.

REFERENCE

1. http://www.nano.gov/html/facts/whatIsNano.html.

Chapter 10

Entrepreneurs in the Technological Ecosystem

Jeff Lawrence and Larry Bock

There are thousands of people with ideas. There are thousands of people who can lead and manage. There are thousands of people who can get things done. There are far fewer people with the unique combination of these qualities that can be labeled as entrepreneurs.

Entrepreneurs aren't satisfied when they can see a vision and express it; they will be satisfied only when that vision becomes real. Entrepreneurs exist in all areas and at all scales of human endeavor—a local business on a street corner, a large technology business spread across the world, or an organization trying to bring about social, political, or economic change. Entrepreneurs become possessed by their vision and will persevere in developing, refining, and spreading that vision until it has changed the world in some way, large or small.

There are many theories, stories, and how-to descriptions of entrepreneurship. This chapter briefly discusses a framework for thinking about entrepreneurship, the role of entrepreneurs, who they are, what they do, what they can expect, and why they do what they do.

LIONS, TIGERS, AND BEARS: WHAT IS THE ROLE OF AN ENTREPRENEUR?

Human and technological progress is both the cause and the effect of a rich and complex set of interlocking relationships and cycles that is similar to many aspects of natural ecosystems. Natural ecosystems consist of small and large thermodynamic, chemical, and biological loops that are necessary to sustain, proliferate, and advance life. By the time an ecosystem matures, either on a local, regional, or even global scale, there are diverse species, stable populations, and complex interdependencies among the members of the ecosystem's communities.

The technology ecosystem is very similar. The loops necessary to sustain an ecosystem consist of people, information, technology, and capital. Capital (that is, money), although very real, is also an abstract method for transferring energy (in the form of work and labor) within and between communities. The food chain (also known as the food web because of a potentially large number of interconnections) describes the various populations and their relationships to each other. In nanotechnology the food chain consists of users, businesses (vendors of technology, products, and information), investors, government, academia, industry forums, and standards organizations.

Each population is trying to sustain itself and proliferate in the context of the development stage and available resources of its community and ecosystem. Ecosystems develop and become populated over time by different populations. Some ecosystem populations reach maturity, but others never fully develop or fail. Occasionally populations fail because of natural disasters or severe disruptions (a meteorite, a forest fire, a stock market crash).

Businesses' and industries' development progress along the population curve is also typically affected by their movement along the hype curve. The vertical axis of the hype curve is visibility, and the horizontal axis is time, divided into periods including the start, the peak of inflated expectations, the trough of disillusionment, the slope of enlightenment, and the plateau of productivity. To some extent, how well businesses and industries negotiate the hype curve will determine whether the population of the ecosystem follows an S-curve to maturity or a J-curve to failure and removal from the gene pool. The early stages of a developing ecosystem are sparsely populated and typically have low species diversity, low inertia (resistance to change), and high resilience (the ability to restore itself after a disturbance). As an ecosystem develops, species diversity and inertia increase and resilience decreases.

Interestingly, some people don't consider nanotechnology an industry per se but rather a series of technologies and products that are applied to solve a wide and possibly unrelated range of problems. In many cases it is unclear whether business synergies can be developed between the disparate uses of the technologies and products. Some of these technologies and products include materials, medical devices, other devices, and processes and tools.

Materials take a long time to get to the market and replace or supplement existing industrial process flows. Planning, procurement, and testing cycles can be very long and are usually tied to capital budgets. Medical devices are expensive and take a long time to get to market because of scientific peer review, government testing, and the need to resolve potential reimbursement issues. Other devices may be less expensive or quicker to get to market, but the value proposition may be more difficult to define.

In all cases, success will depend on having a clear value proposition (in the form of lives saved, quality of life improved, costs or expenses saved, or markets opened) and having sufficient capital available to ensure survival until product sales can scale up. In the early stages of the ecosystem, the development of infrastructure and tools to support the rest of the ecosystem may offer the best possibilities for success.

Nanotechnology resides primarily in academic institutions, federal laboratories, large corporations, and to a lesser extent in small entrepreneurial businesses. There is significant research and some development going on. Nanotechnology is people- and capital-intensive. The tools, processes, and techniques to monitor and control the structural, electronic, optical, and magnetic properties on the atomic scale are expensive. Capital to fund these activities is coming from taxpayers in the form of federal research grants and to a lesser extent from large corporation research and development budgets.

There is some private equity available. The private equity market, in the form of angel investors or venture capitalists, has fairly high expectations that must be met to engage its interest:

- A strong, transparent, predictable, and ethical management team
- A large, identifiable, addressable market
- Good product or service value and a strong, defensible market position
- Strong, growing, and consistent revenue and earnings performance
- An understandable story and strategy, leading to a future liquidity event (in the form of acquisition, buyout, or public financing)

There is still disagreement about whether the prodigious sums of capital that were available for the Internet, telecommunications, and pharmaceutical industries will become available for the nanotechnology industry. Many investors feel that public equity funding will become available as privately or self-funded companies mature and show the predictability and financial performance the public markets expect.

Private equity is not the only way to fund a start-up. Self-funding is possible, but very difficult, because of the capital-intensive requirements of nanotechnology. It can take millions of dollars to perform nanotechnology research and development, although it may not be as capital-intensive as the semiconductor industry on the back end. Licensing or using open source technology or products, rather than creating them from scratch, and using shared labs, facilities, and tools can make self-funding more viable.

The nanotechnology ecosystem has analogous models in the information technology, pharmaceutical, and energy industry ecosystems. These industries can be looked to as examples of how nanotechnology might evolve over time in terms of the landscape, the food chain, the ways they operate in regulated or unregulated environments, capital requirements, infrastructure requirements, innovation and development models, and paths to liquidity. Survival and proliferation through all the stages of nanotechnology ecosystem development will depend on the entrepreneur's ability to create or fill a niche that provides a unique value and becomes an integral part of the relationships and loops of the ecosystem.

As the ecosystem develops, as it has in analogous industries, adaptability and flexibility will be essential for proliferation and advancement. Interestingly, there is a fundamental difference in the ways adaptability and flexibility are realized in nature and in industry. In nature, members of the food chain "float" around, but in industry, entrepreneurs can make deliberate choices of how to adapt, stay flexible, and innovate. The ability to make choices and innovate is important because in the early days of most ecosystem (industry) development, the fast rate of change disfavors large, slow-moving organisms (organizations) and favors organisms (organizations) that are smaller, more adaptable, and responsive. Large organizations in general don't innovate as well as small ones do—not necessarily because they lack the vision, but because risk taking is judged and supported differently than in small organizations.

THE POWER OF AN IDEA: WHAT IS A GOOD IDEA?

All organizations, institutions, and creations of humans have started with an idea. The power of a good idea cannot be underestimated. An idea by itself is a necessary, but not sufficient, condition to successfully effect change, develop a technology or product, or start a business. Many ideas have not gone beyond the stage of being only ideas.

To have an impact, the idea must be wrapped in a vision and an understanding of whom the vision serves. This is where entrepreneurs enter the picture. They work with an idea, either their own or somebody else's, and mold it into a vision that is part possibility (what can be) and part reality (what needs to be). Entrepreneurs power the vision into reality by spending their time and energy and, most importantly, by maintaining their belief about the importance of the vision. Entrepreneurs may find it challenging to keep people focused on the entrepreneurial vision over the long time horizons of nanotechnology.

For many entrepreneurs, ideas come from methodical thinking and planning. Others may stumble upon ideas or find ideas thrust upon them by a life event or epiphany. Good ideas address a real problem or need for change, are realizable, are easy to communicate, and provide understandable value. In many cases, good ideas (such as sticky notes and penicillin) have come from what were perceived as problems by some but were seen as solutions by others.

Ideas can be sustaining or disruptive. Sustaining ideas improve the performance or capabilities of existing technologies or products along dimensions that have been previously valued by customers. They tend to be cumulative in their effect. Disruptive technologies bring to market a different value proposition than was previously available. They tend to be revolutionary and, at least initially, must be sufficiently resilient and flexible to survive ambiguity, incoherence, and uncertainty as the relationships and loops of the ecosystem rearrange themselves to support the new value proposition. In the beginning, successful disruptive ideas are embraced by a few but, as their value proposition becomes obvious, are adopted by many.

Importantly, even if an idea is good, to be successful its timing must also be good. Many ideas coming from academics must be tempered by the realities of business. Educating others about new ideas may take significant time and effort. Nanotechnology as an industry is still fairly new, and although many are interested in its possibilities, as yet few totally understand the technology, its application, and its value in the ecosystem. Some ideas may simply occur too early in the ecosystem's development to be successful. It is up to the entrepreneur to be keenly aware of the state of the ecosystem and to know whether or not it is ready to accept and embrace the idea. If the ecosystem is not ready for the idea, either the idea must be shelved or the ecosystem must be prepared for its adoption.

Entrepreneurs are generally not ideologists; they are pragmatists and innovators at heart. They enjoy seeing their technology, product, or service established in the ecosystem, and they understand that the value of their vision must be unique, real, and understood if it is to be adopted.

A SINGLE PERSON CAN CHANGE THE WORLD: WHAT IS AN ENTREPRENEUR?

A vision doesn't exist by itself. It is a highly personal view of a possible future (what can be). The literature is filled with stories of successful entrepreneurs. Every vision, and the story behind it, is as unique as the individual entrepreneur. Stories serve to show possibilities and to focus and inspire. Although each story is unique, the commonalities in the motivations, risks, and challenges are many.

Entrepreneurs are not born, they are made—sometimes consciously and many times unconsciously. Their path to success tends to go through three phases: learning, epiphany, and doing. The learning phase is usually a lifelong process. A college degree in business may be helpful, but history shows that it is not necessarily required for someone to be successful. Knowing the theories and practices of business, technology, and other disciplines is important, but a would-be entrepreneur may need to develop and refine many other skills—such as listening, learning, judging, deciding, and communicating—through personal experience. Entrepreneurs, more or less intuitively, put themselves through a long apprenticeship during which they develop the knowledge, experience, and skills they will need in the future. To some extent, entrepreneurship is about constantly preparing yourself to deal with the opportunities and challenges that may cross your path in the future.

During the learning phase, entrepreneurs may be building toward an epiphany or may experience a jolt from a completely unexpected source or direction that pushes them out of their comfort zone. The epiphany may be incremental or dramatic. The epiphany is the defining moment, when a new idea or a life event forces a change or rethinking of the entrepreneur's path in life. It is a crystallization of the inklings, intuitions, desires, beliefs, and learning leading up to that moment.

Put simply, from the epiphany forward, entrepreneurs feel and believe that they must pursue the path they have seen for their vision and themselves. They are driven to have an impact on technology, industry, society, or individuals; to gain new knowledge and skills; to make money; to control their own destiny; to prove they can do it; or to gain recognition—or all of these.

Having an epiphany is one thing; making it real is something else. Getting from the first point to the second requires extraordinary persistence and entails numerous risks and uncertainties. Entrepreneurs generally see opportunity where others might see risk or uncertainty. There is usually greater opportunity associated with greater risk and uncertainty.

Risk is relative to the person judging the risk, the situation itself, and whether the person is looking forward or backward in time. Entrepreneurs manage risk and uncertainty by understanding their goals, the environment, and the players, as well as the worldviews, languages, perspectives, intentions, and motivations of possibly affected parties. Entrepreneurs recognize what they don't understand and their ability or inability to position, influence, or control the situation. They judge the alternatives and scenarios against their own and others' experience, knowledge, information, principles, and goals, and they decide on action or inaction and adapt as necessary.

Engineers and scientists are often given to amassing data before making a decision. Entrepreneurs realize that it can be dangerous to avoid making a decision because of a lack of information. Sometimes the information is simply not available or nonexistent, and they learn to work with what information, opinions, and views are available at the time to reach decisions and take actions.

LOOK FORWARD, THINK BACKWARD: WHAT DOES AN ENTREPRENEUR DO?

To be realized, a vision requires time, energy, capital, belief, and drive. It is not enough to talk; it is as important to lead and do. Some people look at where they are and what they have in order to see where they can go (that is, they look forward and think forward). Others see where they want to go and look at what they need to get there (they look forward and think backward).

The two approaches sound similar but in fact are very different. The former is limited to using whatever resources and information are available, whereas the latter doesn't accept this limitation and determines what needs to become available (that is, enabling) to realize the vision. The practice of looking forward and thinking backward will identify the enablers and changes needed for future success. Once identified, these enablers and changes must be encouraged and developed.

Incremental visions can be realized with either approach, but more disruptive visions usually are realized by looking forward and thinking backward. Disruptive visions may require significant changes to the ecosystem before they can be realized. A handful of individual grains of sand will have very little impact when it is thrown against something. But when that same handful of sand is molded into a ball, it will have a much greater impact. Entrepreneurial leadership is about molding the individual grains of sand into a coherent whole. Entrepreneurs lead by

- Giving a voice to the vision and ideas
- Building and inculturating a team or organization and giving it purpose
- Identifying and acquiring resources (people, capital, information)
- Removing obstacles and solving problems

These activities should be performed in the context of a set of values and principles that guide the entrepreneur's behavior. The values and principles vary from entrepreneur to entrepreneur and should represent his or her views about how to develop, manage, and value relationships with employees, customers, investors, and others, as well as the entrepreneur's views about what success is and how to create the conditions necessary for success. These values and principles do not necessarily have to be articulated in detail, but at a minimum they should show themselves consistently in the everyday actions and behaviors of the entrepreneur. Transparency and consistency will allow others to understand the motivations and expectations as well as the likely directions and actions of the entrepreneur, which in turn will allow them to act in concert and contribute to the entrepreneur's vision.

The entrepreneur must understand and carefully consider all aspects of the vision and the steps it must be moved through to become real. Entrepreneurs must constantly deal with and move back and forth between broad, general issues and mundane, detailed issues. Building a company has many challenges, including these:

- Creating value through technologies, products, services, or content
- Measuring value in its tangible, intangible, financial, and strategic forms
- Protecting value using trade secrets, copyrights, patents, trademarks, agreements, and contracts
- Positioning the products, services, content, and company in the ecosystem
- Building and managing relationships with customers, employees, shareholders, advisers, media, analysts, and suppliers

The day-to-day activities of addressing these challenges consist of listening, learning, judging, deciding, and communicating. Entrepreneurs are judged by their ability to meet these challenges, their own expectations, and the expectations of others.

THE GOOD, BAD, AND UGLY: WHAT SHOULD AN ENTREPRENEUR EXPECT?

Entrepreneurs generally commit themselves full-time to realizing their vision. If they are unwilling to make this commitment, then they probably haven't developed the vision to the point where they are confident that it will be realized—or else they are not entrepreneurs. True entrepreneurs cannot rest until their ideas have succeeded.

During the journey from idea to success, entrepreneurs will experience many feelings and emotions. The hours are long and the personal commitment is usually complete, especially in the early stages of the vision's development. It is not unusual for the entire journey to take many years.

Entrepreneurs wake up in the morning thinking about their vision and go to sleep at night thinking about their vision. Work, play, and family blur together. The single-mindedness of purpose an entrepreneur feels is hard to explain to others. A supportive, patient, and understanding family and social network are helpful. Finding somebody to talk to who is familiar with the entrepreneur's vision, but without a personal stake in it, is also helpful as a means to test, practice, and refine thinking and communications skills.

The best idea and vision in the world will not be realized if others can't understand it and can't become excited by it. Sometimes the vision is too complicated to explain easily, sometimes the markets do not exist or are not fully developed, and sometimes institutions and organizations are not ready for the changes that are necessary for the vision to be realized.

It is important for entrepreneurs to understand the worldview and language of the audience they are trying to influence. Scientists, engineers, finance people, salespeople, and marketing people all have very different mental models and views of how the world operates and behaves. These differences manifest themselves in many forms, particularly in language. Entrepreneurs must use this knowledge and make an effort to communicate using the target audience's own worldview and language, which are not necessarily the entrepreneur's own, to influence those things that need to be influenced. Indeed, an entrepreneur must be prepared, especially at the leading edge, to create and teach a new vocabulary so that the vision can be understood and embraced by the target audience.

Entrepreneurs generally lack some or all of the information, knowledge, experience, and skills they need to realize their vision. Entrepreneurs learn very quickly—if they don't know it already—that they can't do everything by themselves. A large part of entrepreneurship consists of on-the-job training and building a team to do the things entrepreneurs can't do themselves.

As they build their team, an interesting question entrepreneurs in nanotechnology must consider is whether they should use proven experts from other fields or novices from nanotechnology itself. The former may bring experience to the challenge, but the latter may bring energy. Time and windows of opportunity are usually short and moving targets, so points of view must be established and decisions usually must be made under uncertain and ambiguous circumstances. There are high degrees of uncertainty, risk, and ambiguity in trying to realize a vision, and it is important to recognize and accept that mistakes will be made. If they aren't made, there won't be progress. Sometimes taking an informed, calculated risk is the best thing an entrepreneur can do. It is also important not to take false comfort in numbers, because they can be manipulated to represent or support almost any point of view.

The world of large companies is very different from the world of small companies. Resources (people, capital, and support), systems, awareness, and reputation can vary dramatically. The transition between these worlds, although not impossible, can be difficult. Large companies offer an opportunity to learn and make mistakes in a relatively safe environment. They have already established themselves as part of the ecosystem, and their survival is usually not an immediate concern. Small companies, on the other hand, tend to be focused on their survival and on establishing a niche in the ecosystem. This means that day-to-day operational issues take priority over more strategic or long-range issues. Small companies have the advantage that their small size and organizational looseness provide them with adaptability and flexibility to move through the ecosystem in ways a large company cannot.

As a practical matter, whether self-financed or investor-financed, entrepreneurs can expect, at best, to sustain their existing lifestyle. The financial rewards of pursuing your vision, if they are ever realized, probably won't come for many years in the future through some form of liquidity event such as an acquisition, public financing, or buyout.

At its heart, any successful business depends on its people. People and their behaviors are not always logical or predictable. Engineers and scientists sometimes have a difficult time dealing with this. During the development of a business an entrepreneur will experience the full breadth of human behavior. Small organizations tend to be self-selecting and usually don't represent the entire range of human behavior. As an organization grows, the odds are that a broader range of human behaviors will be encountered and in some instances will be surprising and challenging.

Delegating responsibility and authority and letting others do things their own way is also a difficult transition for many entrepreneurs. Once they get through this transition it can be liberating, because they are able to give up tasks to others and free themselves to do new and interesting things. They must understand their own and evolving value and niche within the ecosystem of their organization and anticipate how it will change over time.

Moving a vision forward is fraught with skepticism, criticism, conflict, surprises, and mistakes. Entrepreneurs must listen to skepticism or criticism openly and not take it personally, and they must understand, accept, and deal with any truths that may have been expressed, no matter how ugly they might be. It is important to learn from adversity and delight in the surprises. Confidence in your ability will develop over time.

Leadership in its simplest form is about getting people somewhere they don't know they need to be. Inertia will be a strong counterforce to change, and the act of overcoming it may take large amounts of time and energy.

There will be a constant need and opportunity to change and adapt to new situations.

IS IT WORTH IT?

Many people start down the path of entrepreneurship, but only a few are successful at fully realizing their vision and other aspirations. Entrepreneurs are driven by the chance to make an impact, to learn, to prove something to themselves or others, or to make money. All entrepreneurs have a unique definition of success and a list of expectations for themselves and others.

Successful entrepreneurs have a vision, understand whom they serve, understand the vision's place in the ecosystem, and give purpose to the organization they build. They must throw themselves wholeheartedly into realizing their vision and must be comfortable with risk, uncertainty, and ambiguity; must be good listeners and learners and strong communicators; and must be adaptable and flexible and pragmatic problem solvers. The necessary experience and skills can be developed on the job and by learning from others.

Even though entrepreneurs can control and influence many things, a certain amount of good luck is also required to be successful. One cannot underestimate the importance of being at the right place at the right time or meeting the right person. An entrepreneur's days can be long, stressful, and tiring, but they can also be exhilarating and joyful. The team building and the shared struggle of trying to realize a vision can be intense and satisfying. The potential rewards can be huge.

Entrepreneurship is a unique and privileged opportunity to pursue a vision. Even though the challenges are daunting at times, true entrepreneurs would say that their experience, whether or not they were successful, was worth it, and in fact it would be hard for them to imagine having done anything else in their lives.

Chapter 11

Major Corporations: Technology, Business, and the Culture of Opportunity

Jim Duncan

Your interest in this book is likely based on one of two reasons: You are a technologist looking for insight in how to build a business, or you are from the business side and are trying to grow a company from the technology.

This is the usual paradox of any technology-based venture and doubtless an extreme challenge when nanotechnology is involved. As this book is being composed, the first of a new generation of nanotechnology-enabled products are finding their way to the market, and plenty more are being developed. Perhaps an even greater effort is being made to develop the tools and processes that many future generations of products will be based on.

Although it may be impossible to address the specific issues in the evolution from nanoscience to nanoproducts, we can address some of the classic issues surrounding technology-based opportunities.

CULTURE, MODELS, AND CYCLES

There is enormous variety in the attempts to manage these opportunities from both a technology and a business perspective. Each has its strengths and weaknesses. The prospects of success are better with an integrated approach, with success more probable when an integrated team addresses development issues.

It is also crucial to use management processes that are appropriate for the technology. Processes developed for mature products are rarely effective for emerging and new technologies. Despite the definition and control the scientific method brings, innovation remains a creative process. Outside of independent government-funded research centers and fewer still commercial centers, research centers strive to make their technology relevant to present and future problems.

On the one hand, few research and development organizations can afford to explore science and technology without an eye to its commercial potential. The potentially huge return on investment creates the business motivation to provide the resources for the exploration. The "best" of these ideas are those that are thought to have the best commercial potential. In theory, the best ideas get the necessary resources. On the other hand, not many businesses can afford the sustained investment required to examine the science and explore the technology that results in a successful product and the protective set of enabling intellectual property.

Accordingly, the business needs and the state of the technology have an entwined relationship.

In the technology-based community, conversation often centers on the technology, its promises, and its challenges. We often include the myriad possible applications as a validation of the effort. More often than not, the applications and eventual product are our dreams. We quietly acknowledge that these pursuits require different skills. The possibilities are, however, the stuff that fuels our imagination and motivates our investigations of science and new technology.

In parallel, the business community hopes to have the technology that will give it a sustainable, ethical advantage for its product in the marketplace. As business executives look toward the research and development community, while managing the profit and loss expectations of investors and owners, it is a challenge for them to nurture the technology, apply it, develop a product, and market it. Anyone that has managed or led technology from its R&D origins to a fully developed product knows that the process has a staggering amount and variety of variables.

The process can be reduced to two major stages: research and development and commercial enterprise.

The research and development stage has two major subcategories: science and technology. Science is the way we learn about the physical universe by applying the principles of the scientific method: observation, hypothesis, and testing. Technology is the application of the science to develop tools, techniques, materials, and systems.

The commercial enterprise stage has three major subcategories: application, product, and market. In the application phase, the technology is brought to bear to solve a customer problem. Product is the commodity that is for sale, and marketing identifies customers for those commodities.

The two major areas—research and development and commercial enterprise—are cultures apart. They are separated by a gulf.

The gulf is the boundary for the two points of view: technology and business. The technical mind is trained to believe and follow the scientific method, and scientific truths and laws govern the process. The business mind is not very different, but there are vast market forces that cannot be controlled or isolated and can be serendipitous. It is those forces that determine the market value of an opportunity.

More often than not, the research and commercial enterprises exist as separate organizations. Only a few of the largest companies have resident R&D staff. Small, medium, and even some large companies often rely on outside sources for technology. Whether resident or not, they are separated by the gulf of cultural differences. Language barriers, differing goals, and differing metrics for success define the gulf. The challenges of the commercial enterprise deserve as much consideration as those of the research enterprise. Therefore, we are addressing the approach to the gulf from the technology side.

Bridging the gulf is *the* challenge of developing technology into real products. In other words, the opportunity cannot be only about the technology, nor can it be only about the business. The technology must serve the business, and the business must respect and nurture the technology. Commercial enterprises are concerned about the role technology plays in their competitive position. A general assumption is that a strong intellectual property position is a tool to protect profit margin. Profit margin is only one of the important business metrics of a flourishing commercial enterprise.

Both technology and business enterprises must stretch across their boundaries and reach for inspiration and guidance.

THE HOLY GRAIL

The greatest discovery of them all may be the process that guides us through the evolution of science to product, especially one that is capable of taking advantage of a great business opportunity. Our hopes in this regard form on two points of view: the science and the business.

From the technology point of view there is a tendency to hope that after the application of the technology to an existing problem, a business will pull the technology forward through the remaining commercialization steps. From the business point of view there is a hope that the technology will present itself as a product and will need only to be marketed and sold. Neither eventuality is likely.

The chances of success are increased when you have an appreciation of the difference and culture of each of the environments. The technology needs to be mentored by an understanding of the anticipated needs of the business, and the business must be vigilant of opportunities presented by emerging technology.

The ultimate process is more of an approach. The approach is both flexible and adaptive. The effort involves multidisciplinary teams of creative, success-driven professionals who have entrepreneurial talent. It's obvious when one is in the presence of a process that is working to take the technology out of the lab environment and into the marketplace. It's evident in the team members' relationships among themselves and their relationship with organizational management. The evidence is a mutual respect and appreciation for each of the necessary contributions to the opportunity.

The process is the development of a culture of innovation, irrespective of whether the parent organization is a commercial or a research enterprise.

The challenge in either situation centers on a single core concept: The opportunity must be managed by using processes that are appropriate for the maturity level of the technology. For emerging technology there must be room for error, failure, reevaluation, rehabilitation, and even abandonment. However, being afforded the opportunity to take the challenge should be seen as a high compliment.

The professionals involved and leading these efforts must exhibit a number of special attributes. Look for the presence of most of the following eight points in every member involved in the effort to bring the technology forward:

- **Confidence.** Confidence leverages training, qualifications, and experience for the individual to take calculated but not reckless risks.
- **Focus.** Participants must be outcome oriented.
- **Self-sufficiency.** The ability to operate and thrive independent of broad and deep organizational infrastructure stimulates creative problem solving and obstacle removal.
- **Adaptability.** Adaptability is active learning: analyzing, solving, and adjusting.
- **Emotional stability.** The desired team member balances the range of emotions (fear, anger, frustration, excitement) and does not reject the extremes for the median.
- **Insightful.** Being insightful in this case is the ability to know what is important in a given situation, to know what it takes to get a task done, and to filter distractions.

- **Courage.** Life and entrepreneurial ventures are not without their obstacles of all natures. The team member must be able to face these challenges.
- **Motivation.** The team wants its members to be appropriately inspired to start the journey and continue on the path.

The culture of opportunity balances soft and hard points. The culture addresses the needs and characteristics of the people, the technology, and the business. The balancing starts at the gulf where research and commercial enterprises meet.

Chapter 12

Nanotechnology in Federal Labs

Meyya Meyyappan

Federal laboratories play a major role in the development of nanotechnology. Some of the key players include NASA's Ames Research Center, U.S. Naval Research Laboratory, Army Research Laboratories, the National Institute of Standards and Technology (NIST), and the various Department of Energy (DOE) national laboratories around the country. The focus of the efforts in these laboratories is technology development for the needs of the parent organization.

For example, the impetus for nanotechnology research at NASA labs is the development of advanced miniaturization of sensors and instrumentation for science missions to meet the need for smaller mass, volume, and power consumption. Some of the mission needs expected to benefit from nanotechnology include the development of radiation-hardened electronics and communication devices, microcraft (weighing only 10kg or less), high-strength but low-weight composites for future spacecraft, and a sensor network consisting of thousands of physical, chemical, and biosensors for planetary exploration.

Federal laboratories in the Department of Defense are focused on developing ultrasensitive and highly discriminating sensors for chemical, biological, and nuclear threats. Revolutions in electronics, optoelectronics, and photonics devices for gathering, protection, and transmission of information are critical for DoD's missions. The goal of Army Research Lab in its nanotechnology efforts is to reduce the weight carried by individual soldiers without losing any functionality. DOE national laboratories serve the agency's missions related to both basic and applied energy research as well as national security. The DOE national laboratories also include the nation's largest network of user facilities, open to all researchers on a peer-review basis, including five Nanoscale Science Research Centers.

THE ROLE OF FEDERAL RESEARCH LABORATORIES

As research and development components of participating agencies of the National Nanotechnology Initiative (NNI), the federal laboratories address various grand challenges in nanotechnology, including nanomaterials for energy needs, metrology, nanoelectronics, materials for chemical industries, health-care needs, and space exploration.

The broader goals of many federal agencies are also addressed by academic universities through contracts and grants, and by small businesses through research contracts from the Small Business Innovation Research (SBIR) and Small Business Technology Transfer (STTR) programs. All these augment the in-house efforts within the federal laboratories.

In addition to serving the needs of the federal labs' parent organizations, the devices, processes, and technologies developed at federal labs often have commercial applications. Federal laboratories have historically been a rich source of new commercial products, with hundreds of successful stories of tech transfer to the private sector over the past forty years. Products such as cordless drills emerged from the Apollo program because of the need for astronauts to drill down beneath the moon's surface and use compact, battery-powered tools. Wilson Sporting Goods Company used technology developed for the Space Shuttle's tanks to create a more symmetrical golf ball surface. The resulting selection and placement of dimples on the surface of the golf ball help the ball travel longer distances. The Space Shuttle's thermal protection technology is now used to insulate race cars from sweltering interior temperatures. Technology originally developed for NASA's Skylab led to the home smoke detectors commonly used today. Research at DOE national laboratories has contributed to the development of an artificial retina that uses a nanocrystalline diamond film and spun out a company that is one of the leading producers of specialized nanoscale materials.

These are only a few examples of commercial products that can trace their origin to one of the federal labs. One can expect this trend to continue with nanotechnology as well in the coming years. There is a well-established mechanism for technology transfer to the commercial sector. In the case of NASA, there is the Office of Technology Partnerships, which works on all the technology licensing programs. *NASA Tech Briefs* magazine, a monthly publication, lists and describes NASA innovations that are ready for the marketplace.

In addition, NASA has regional tech transfer offices that focus on matching a particular innovation with the needs of the industry in the region. Regular tech transfer conferences are also organized by the field centers, and regional tech transfer offices bring together NASA innovators and industrial participants. All possible avenues are used to get the word out about the availability of technologies for commercial licensing.

As an example of federal laboratory activities, consider technology transfer efforts at one of the leading nanotechnology laboratories at NASA Ames Research Center. The research focus at NASA Ames is on nanomaterials such as carbon nanotubes, inorganic nanowires, conducting organic molecules, and protein nanotubes. The applications focus for these nanomaterials is diverse: nanoelectronics, computing, data storage, nanoscale lasers, chemical sensors and biosensors, ultraviolet and infrared detectors, and advanced life-support systems that address waste remediation, air purification, and water purification as well as instrumentation for planetary exploration and astronaut health monitoring devices.

TRANSFERRING TECHNOLOGY

NASA Ames has developed a substantial intellectual property portfolio in these areas. Some of the technologies have been licensed to private companies. For example, the use of carbon nanotubes in computer chip cooling and associated processing techniques have been licensed to a start-up company in the Bay Area. Similar efforts are ongoing for licensing biosensors, genechips, and chemical sensors. Another avenue of tech transfer has been companies founded by employees of the NASA Ames Center for Nanotechnology.

NASA Ames Office of Technology Partnerships routinely organizes one-day workshops to showcase the Ames-developed technologies. To meet entrepreneurs and members of the industrial community, this office also participates in such workshops organized by other NASA centers and regional tech transfer centers, as well as other conferences and symposia.

In addition to working on tech transfer to industry, this office facilitates joint research between NASA Ames and interested companies through an arrangement called the Space Act Agreement (SAA). The SAA is a type of cooperative research and development agreement that enables the two parties to jointly develop a process or product, with each contributing to certain aspects of the development. The rights of each party for commercialization and marketing are clearly spelled out in the SAA document as negotiated by the two parties. NASA Ames currently has several SAAs in place with a number of industrial partners.

Technology transfer in Department of Defense labs takes place in a variety of ways. The primary focus of DoD labs is not to develop commercial products but to develop solutions to the problems of military services such as the Army, Navy, and so on. These services communicate problems, issues, and limitations they experience in the conduct of their operations, and priorities are developed by the DoD labs to resolve these problems. Solutions are proposed by the labs through written and oral proposals to meet specific needs, with funding provided (when available) to those of the highest priority.

For additional research and scientific validation, labs may also partner with an outside company or university through a Cooperative Research and Development Agreement (CRADA). This can leverage both the funding and the expertise of the laboratory to greater meet the needs of the military services. When a project has reached this point and is close to developing a prototype, funding may be obtained from DoD sources outside the lab, such as a branch's Scientific Office of Research or funding from an applied program (for example, the F-22 aircraft or global positioning satellites). Defense labs rarely directly spin out commercial companies, but they have extensive commercialization and licensing programs similar to those of other agencies.

DOE's Nanoscale Science Research Centers convene regular meetings to provide a forum for exchange of ideas among researchers from many institutions, as well as a means to educate users on the capabilities that exist and are being developed at these unique national facilities. The facilities are available for use on a peer-reviewed basis. Time, materials, and technical assistance are provided free of charge for researchers who are willing to publish the results of their work in the open literature, and on a cost-recovery basis when used for proprietary purposes.

SUMMARY

There are active nanotechnology programs with a substantial investment of personnel at many federal laboratories. These mission-driven labs explore specific technologies, seeking solutions to urgent and long-standing problems in their road maps. Invariably the developments have commercial possibilities, and each lab has a commercialization office and well-established procedures to transfer technology to industry. In addition to the nation's universities, the federal labs are anticipated to be a key source for intellectual property in nanotechnology.

SECTION THREE

Materials and Industries

Chapter 13

Nanoscale Materials

OVERVIEW

Mark Reed

Materials form the basic underlying building block of nearly every advanced technology, and nanotechnology is no exception. Most technologies have various levels of differentiation, which can be broken down roughly into (1) materials, (2) design and fabrication, and (3) integration. In the case of some conceptual nanotechnologies, the path to achieve the last two levels is not at all clear. However, the science and technology of nanomaterials are both vibrant fields, with breathtaking advances on a daily basis. In many ways contemporary nanoscience is essentially the study of nanomaterials.

What differentiates nanomaterials from classical materials science, chemistry, and the like is the degree of control. Whereas many scientific disciplines have for more than a century dealt with phenomena and understanding on the atomic scale, the emerging field of nanomaterials strives to achieve control on that level beyond stochastic processes. One might view this field as a variant of engineering; instead of studying what is, we try to create what never was—and now on an unprecedented level.

One of the original areas of nanoscience—the production of nanometer-scale particles of various materials—is nearly as old as human civilization. Not until the past few decades have researchers begun to appreciate the unique properties of nanoparticles. With the understanding of structure and synthesis, researchers worldwide have gained nearly atomistic control over size and properties, creating such intriguing structures as "core-shell" quantum dots. Nanoparticles are finding applications in a number of diagnostic and biomedical uses and will become widespread when toxicity and functionalization issues are solved.

The unarguable poster child of nanoscience is the carbon nanotube (CNT), which are graphene sheets rolled into a perfect cylinder with mind-bending aspect ratios, often being only a nanometer in diameter but many micrometers in length. More interestingly, their electronic structure is dramatically dependent on dimension and twist—an exciting advantage *if* these parameters can be controlled, but generally viewed as a limitation *until* these problems can be solved. Nonetheless, fascinating electrical, thermal, and structural properties of single-walled nanotubes (SWNTs) have been measured. The scaling of this understanding to large-scale synthesis and the use of these properties on a macroscopic scale remain challenges.

More recently, inorganic semiconducting nanowires have begun to attract attention as an alternative to nanotubes because their electronic properties are easier to control. First explored in the early 1990s, these single-crystal nanowhiskers, with dimensions of only tens of nanometers, have seen a resurgence in the past few years because of the development of a wide variety of synthesis methods. In addition, the range of materials that can be fabricated into nanowires is impressive, giving a large design space for property control. Although the exploration of these systems is still in its infancy, this field represents a rapidly expanding frontier of functional nanomaterials.

Finally, the interface between nanomaterials and numerous applications, such as biomedical, requires nonstandard materials. Whereas soft and polymeric materials are not nanomaterials, these materials produce critical fabrication and interface implementations, and an understanding of their properties is essential to modern nanomaterials applications. An extension of this field is to integrate polymeric and single-molecule structures into other nanostructures and complex hybrid materials. These represent the most challenging nanostructures investigated to date.

The field of nanomaterials has a long way to go to reach an understanding and control on the truly atomic scale. A combination of innovative synthesis approaches and new characterization techniques, and an understanding of fluctuations and control at this new length scale, is needed, and it will lead to revolutionary advances in nanomaterials and eventually in nanoscience.

NANOPARTICLES

Sheryl Ehrman

Sometimes called the building blocks of nanotechnology, nanoparticles (particles with diameters less than 100nm) constitute a commercially important sector of the nanotechnology market. Unlike many speculative applications of nanotechnology, nanoparticles have been with us for a while. Nanoscale gold clusters have been used to color ancient glass as far back as Roman civilization.[1] More recently, in the twentieth century, carbon black—a material composed of nanoparticles of high-grade carbon "soot"—was incorporated into tires, resulting in greatly improved durability. By the year 2000, carbon black for tires was a 6-million-tons-per-year global market.[2] Nanoparticles are useful when their properties at the nanoscale (mechanical, optical, magnetic, and so on) are different from those at the bulk in some beneficial way and also when their size enables interactions with biological systems.

An example nanoparticle is shown in Figure 13–1.

Applications of Nanoparticles

Remember white, pasty sunscreen? It is now transparent because the key ingredient, zinc oxide, which absorbs ultraviolet light, is transparent if it is made of nanoparticles but white when made of larger, micron-sized particles. The same holds for titanium dioxide—which is a white paint pigment and an additive to improve opacity in colored paints—if the particles are hundreds of nanometers in diameter or greater. Now, these transparent nanoparticles, which still absorb in the UV range, are finding their way into modern, transparent, and highly effective sunscreens.[3]

Nanoparticles of titania are also a key ingredient in dye-sensitized solar cells. The efficiency of these cells is boosted by incorporating nanoparticles that increase the total surface area for harvesting light by a factor of a thousand as compared with a single crystal of titania.[4] Compared with silicon-based photovoltaic materials, dye-sensitized solar cells are much less expensive to produce, and in 2001 a manufacturing plant opened in Australia.[5]

Another significant market for nanoparticles is in the semiconductor industry, in a process known as chemical mechanical planarization (CMP). In the manufacturing of computer chips, the component materials must be

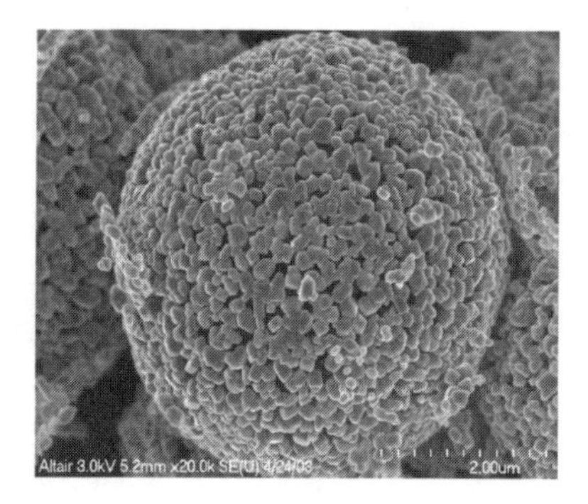

Figure 13–1 TiO2 particles (micron scale formed from nanoscale particles, used for clarity). (Courtesy of Altair Nanotechnologies, Inc.)

coated onto wafers, and at several points in the process, the coatings must be rendered nearly atomically smooth across the entire wafer. For 300-mm wafers, this is no small challenge! In CMP, slurries of nanoparticles are used. In this process, a combination of chemical removal and mechanical abrasion acts to accomplish this atomic-level polishing task. Silica, alumina, and ceria nanoparticles, the most common materials, are very effective. As a result, the global market for CMP has grown rapidly, from $250 million in 1996 to more than $1 billion in 2000, with the market for CMP consumables alone (slurries and polishing pads) expected to hit $800 million by 2005.[6] With ever-increasing miniaturization in device features, resulting in a need for increasing precision in manufacturing, the CMP market will surely continue to grow.

In addition to these existing large-scale applications for nanoparticles, many more products have been recently commercialized or are in development. Professors Paul Alivisatos of the University of California, Berkeley, and Moungi Bawendi of MIT developed processes for making semiconducting nanoparticles out of materials including cadmium selenide (CdS) and cadmium telluride (CdTe). These particles, coated with zinc sulfide, absorb light in the UV range, and then because of size-dependent quantum confinement effects, they emit light in the visible range, at a wavelength dependent on the particle size. Their stability and brightness are much better than those of conventional fluorescent chemical dyes, and when they are functionalized with proteins, oligonucleotides, or smaller molecules of interest, they can be used for fluorescent labeling applications in biotechnology.[7] Silicon nanocrystals less than 4nm in diameter also emit light in the visible range, with the wavelength depending on particle size. These nanoparticles are of interest for solid-state lighting applications and promise much higher efficiency as well as longer lifetimes compared with current incandescent and fluorescent lighting technology.[8]

As well as these optical properties of nanoparticles, chemical properties—in particular, catalytic activity—can be greatly improved at the nanoscale. One important example is gold. In bulk form, gold is relatively inert. However, when deposited at very low concentrations (0.2 to 0.9 atomic %) in a nonmetallic form onto nanoparticles of cerium dioxide, it becomes very active for the water–gas shift reaction.[9] In this reaction, carbon monoxide and water are converted to carbon dioxide and hydrogen. For fuel cells that are powered by hydrocarbon fuels, converted first to hydrogen and carbon-containing by-products, the water–gas shift reaction is key for maximizing the amount of hydrogen while minimizing the amount of carbon monoxide, a molecule that poisons the electrocatalysts in the fuel cell itself. Because of the small amount of precious metal used in this material, the economics are favorable compared with catalysts that contain precious metals at up to 10% atomic percent concentration.

Magnetic properties of nanoparticles can also be exploited—for example, the tendency of small magnetic nanoparticles to be superparamagnetic. Superparamagnetic nanoparticles, in the absence of a magnetic field, have random magnetic moments at temperatures above their Curie temperature, but in the presence of an external magnetic field, they align with the field, producing a high magnetic moment. One application of this phenomenon is in magnetic resonance imaging (MRI). MRI contrast is aided in the body by the presence of naturally occurring materials such as deoxyhemoglobin but can be greatly enhanced by introduction of superparamagnetic iron oxide nanoparticles (SPIONs). These particles have cores consisting of magnetite (Fe_3O_4), maghemite (gamma Fe_2O_3), or some combination of both, and are coated to enhance colloidal stability and biocompatibility. The type of contrast enhancement enhanced by SPION is useful in the imaging of features that contain large amounts of fluid, such as internal injuries or cancerous lesions. These nanoparticles are commercially available from several sources.[10] The surface of the SPION can be further functionalized to engineer interactions between the contrast agents and specific tissue and cell types, and this is an active area of research.[11]

Production of Nanoparticles

Production methods for nanoparticles can be loosely classified into three general categories: wet synthesis, dry synthesis, and milling. In both wet and dry synthesis, nanoparticles are generally produced in a bottom-up way from atomic precursors, whereas in the milling approach, nanoparticles are produced from the top down by mechanically breaking down larger particles. Wet approaches include sol-gel and precipitation methods, whereas dry approaches encompass combustion, furnace, and plasma synthesis of nanoparticles.

In all cases, there are concerns about the narrowness of the size distribution of the nanoparticles, and concern about the degree of agglomeration. All processes for making nanoparticles lead to some spread in the particle size. The size distribution can be modified somewhat by adjusting the process parameters, or the size distribution can be tailored by removing the tails of the distribution through additional separation steps. This typically leads to lower process yield. With respect to agglomeration, nanoparticles have a high ratio of surface area to volume, and it is much more energetically favorable for them to reduce their surface area by coalescing together. Thus, materials that melt at high temperatures if they are in bulk form may fuse together at much lower temperatures if they are nanoparticles.

Some applications, such as the fluorescing quantum dots mentioned earlier, have very tight requirements for particle size and aggregation. Other applications, such as CMP slurries, may not have such narrow constraints. There is no one perfect process. Milling is very energy-intensive, and it may not work at all for some materials, such as pure metals, that are malleable. Precipitation methods may require the addition of capping ligands to the nanoparticle suspension, to stop particle growth and to prevent the particles from agglomerating together. These ligands bind to the surface of the particles, and if they do not impart the desired functionality to the particles, they must be displaced in a separate processing step.

In high-temperature synthesis of nanoparticles, agglomeration can be avoided for some materials by simultaneously quenching and diluting, but this presents a challenge when the process is scaled up. If the nanoparticles suspended in the gas are more dilute, more energy is required to recover them. For some materials that are viscous glasses at high temperature, such as silica, diluting and quenching may not help. Some processes, such as electrospray or plasma-based syntheses, produce particles that are highly charged, and this aids in reducing agglomeration.

Outlook for Nanoparticles

Before a process can be considered commercially viable, there are additional economic concerns. Many processes for nanoparticles have been developed at the laboratory scale, but they are not yet commercialized because of constraints, including scalability considerations and precursor costs. Aerosol pyrolysis—the process used to make nanoparticles of silica, alumina, and titania by companies such as Cabot, Degussa, and DuPont—has long been in existence. Plasma-based processes, because of their potential for high-throughput reduced agglomeration, have been more recently scaled up in industry.[12] Many other processes could perhaps be scaled up, but there is a catch-22 involved: Without a sufficient commercial market for the nanoparticles, there are insufficient resources available for process development. Without sufficient quantities of nanoparticles available, it is difficult to conduct product development research.

In the past several years, there has been increasing public concern about the occupational and environmental risks associated with nanoparticles.[13] The correlation between particle phase air pollution and adverse health effects has been noted for some time.[14] There is now concern that nanoparticles may also threaten human health and ecosystems. It is important to consider that materials that may not pose health risks in bulk may become toxic in nanoparticle form because of their small size. Safety precautions that are appropriate for materials in bulk form may prove insufficient for nanoparticles. As a result, research into potential health risks associated with these new materials, supported by U.S. government agencies such as the Environmental Protection Agency and the National Science Foundation, is actively under way.

REFERENCES

1. J. S. Murday, "The Coming Revolution: Science and Technology of Nanoscale Structures," *The AMPTIAC Newsletter* 6 (2002): 5–10.
2. A. M. Thayer, "Nanomaterials," *Chemical and Engineering News* 81 (2003): 15–22.
3. M. Oswald and K. Deller, "Nanotechnology in Large Scale: Pyrogenic Oxides—Ready for New," *Elements, Degussa Science Newsletter* 9 (2004): 16–20.
4. B. O'Regan and M. Grätzel, "A Low-Cost, High-Efficiency Solar-Cell Based on Dye-Sensitized Colloidal TiO2 Films," *Nature* 353 (1991): 737–740.
5. http://www.sta.com.au/index.htm.
6. R. K. Singh and R. Bajaj, "Advances in Chemical Mechanical Planarization," *MRS Bulletin* 27 (2002): 743–751.
7. A. P. Alivisatos, "The Use of Nanocrystals in Biological Detection," *Nature Biotech* 22 (2004): 47–52.
8. http://www.innovalight.com/.
9. Q. Fu, H. Saltsburg, and M. Flytzani-Stephanopoulos, "Active nonmetallic Au and Pt species on ceria-based water-gas shift catalysts," *Science* 301 (2003): 935–938; and Q. Fu, W. L. Deng, H. Saltsburg, and M. Flytzani-Stephanopoulos, "Activity and stability of low-content gold-cerium oxide catalysts for the water-gas shift reaction," *Applied Catalysis B—Environmental* 56 (2005): 57–68.
10. S. Mornet, S. Vasseur, F. Grasset, and E. Duguet, "Magnetic nanoparticle design for medical diagnosis and therapy," *Journal of Materials Chemistry* 14 (2004): 2161–2174.
11. A. M. Morawski, G. A. Lanza, and S. A. Wickline, "Targeted contrast agents for magnetic resonance imaging and ultrasound," *Current Opinion in Biotechnology* 16 (2005): 89–92.

12. http://www.nanophase.com/technology/family.asp.

13. R. Weiss, "For Science, Nanotech Poses Big Unknowns," *Washington Post*, February 1, 2004.

14. D. W. Dockery, C. A. Pope, X. P. Xu, J. D. Spengler, J. H. Ware, M. E. Fay, B. G. Ferris, and F. E. Speizer, "An Association between Air-Pollution and Mortality in 6 United States Cities," *New England Journal of Medicine* 329 (1993): 1753–1759.

CARBON NANOTUBES

Brent Segal

Since their discovery in 1991 by Sumio Iijima, carbon nanotubes have fascinated scientists with their extraordinary properties.[1] Carbon nanotubes are often described as a graphene sheet rolled up into the shape of cylinder. To be precise, they are graphene cylinders about 1–2nm in diameter and capped with end-containing pentagonal rings. One would imagine that a new chemical such as this would be discovered by a chemist, slaving away in front of a series of Bunsen burners and highly reactive chemicals, with a sudden epiphany being revealed in a flurry of smoke or precipitation from a bubbling flask. However, carbon nanotubes were discovered by an electron microscopist while examining deposits on the surface of a cathode; he was performing experiments involving the production of fullerenes, or buckyballs.

This discovery presents one of the key tenets of nanotechnology. Novel tools allow researchers to observe materials and properties at the nanoscale that often have existed for hundreds or thousands of years and to exploit the properties of such materials.

After Iijima's fantastic discovery, various methods were exploited to produce carbon nanotubes in sufficient quantities to be further studied. Some of the methods included arc discharge, laser ablation, and chemical vapor deposition (CVD).[2] The general principle of nanotube growth involves producing reactive carbon atoms at a very high temperature; these atoms then accumulate in regular patterns on the surface of metal particles that stabilize the formation of the fullerenes, resulting in a long chain of assembled carbon atoms.

The arc-discharge methodology produced large quantities of multi-walled nanotubes (MWNTs), typically greater than 5nm in diameter, which have multiple carbon shells in a structure resembling that of a Russian doll. In recent years, single-walled nanotubes (SWNTs) using this method also have been grown and have become available in large quantities. The laser ablation method of carbon nanotube growth produced SWNTs of excellent quality but requires high-powered lasers while producing small quantities of material. The CVD method was pioneered by Nobel Laureate Richard Smalley and colleagues at Rice University, whose experience with fullerenes is nothing short of legendary. This growth technique is aided by a wealth of well-known inorganic chemicals specifically involving the formation of highly efficient catalysts of transition metals to produce primarily single-walled nanotubes.

Figure 13-2 shows a simple carbon nanotube.

Novel Properties

Although carbon nanotubes have a suitably interesting structure, there are a multitude of important properties that impart the potential for novel applications of significant commercial value. Multiwalled and single-walled nanotubes have similar properties, and for illustration, focusing on single-walled nanotubes provides a reasonable primer of the primary features.

Some of these properties include remarkable strength, high elasticity, and large thermal conductivity and current density. Several reports have determined that SWNTs have a strength of between 50 and 100 times that of steel.[3] The elasticity of SWNT is 1–1.2 terrapascal (TPa), a measure of the ability of a material to return to its original form after being deformed. Imagine a molecule that, on the atomic scale, is as strong as steel but flexible like a rubber band!

Despite these structural properties, SWNTs have a thermal conductivity almost as great as twice that of diamond, which is known to be one of the best conductors of heat. Perhaps one of the most impressive properties of SWNTs involves their electrical conductivity, which is reported to be 10^9 Amps per square centimeter, which is about 100 times that reported in copper, the conductor of choice for nearly every electrical device in common use today.

SWNTs have two types of structural forms, which impart an additional set of electrical characteristics. Depending upon the alignment of the carbon atoms in the cylindrical form, SWNTs can be either achiral (having atomic

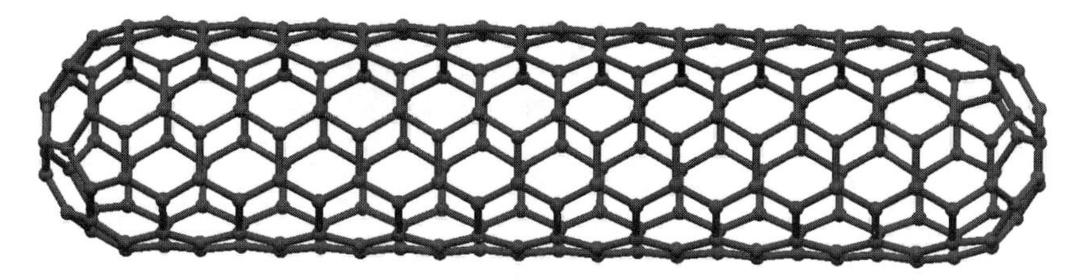

Figure 13–2 A simple example of a carbon nanotube. (Courtesy of Dr. Peter Burke, University of California, Irvine.)

uniformity along its axis) or chiral (having a twisted alignment from the uniform case). Achiral and chiral forms can act as metals or semiconductors and yet retain the same basic nanotube structural motif.

In addition to these well-known properties, SWNTs have other features that make them attractive beyond their scientific novelty. SWNTs have a density approximately half that of aluminum, making them an extremely light material. SWNTs are stable at temperatures up to 2700°C under vacuum. This is truly impressive considering the melting point of Ruthenium, Iridium, and Niobium metals are about the same temperature.[4] Although nanotubes have structural uniformity, the carbon atoms within them have the same precise features as a typical graphene sheet. These atoms can be derivitized to alter the structure of the SWNTs, allowing their properties to be tailored. This allows nanotubes to be subject to literally hundreds of years of rich organic chemistry.

Carbon nanotubes not only can be functionalized to change their structure but also can interact beneficially with organic chemicals that have biological usefulness. Fullerene materials have been explored for use as antioxidants, drug-delivery agents, and amino acid replacements; the latter could lead to new drug candidates. Reports have been published of their benefits in enhancing virus-specific antibody responses or as a scaffold for growing retinal epithelial cells to be transplanted into the retina to treat macular degeneration.[5]

Manufacturing and Scaling Issues

Carbon nanotube synthesis in recent years has been driven by yields and cost. To move nanotubes from scientific curiosity to practicality, they must be available in sufficient quantities at a reasonable cost with high uniformity and reproducibility. In the case of MWNTs, the arc-discharge method provides a good alternative, yielding large quantities of material at a good cost. In the case of SWNTs, while generating large quantities of material, the purity is often unacceptable for a subset of applications because of excessive carbonaceous contamination. Instead, the CVD method and a recent alternative, plasma-enhanced chemical vapor deposition (PECVD), have burst onto the scene as the methods of choice for producing large quantities of SWNTs with micron lengths, purity, and reliability within specifications for certain applications. Figure 13–3 shows an example of PECVD nanotubes.

PECVD has been reported to lower the temperature of nanotube growth significantly by using a plasma to generate the reactive carbon atoms instead of very high temperatures, as in standard CVD growth. In the case of nanoelectronics—perhaps one of the early applications for which SWNTs will

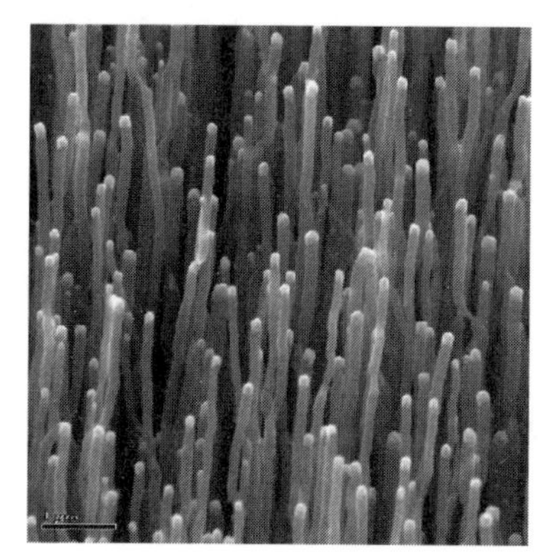

Figure 13–3 Array of carbon nanotubes grown by plasma-enhanced chemical vapor deposition.

be adopted—extremely high purity nanotubes are required, and only a few providers have managed to generate such materials at a reasonable cost.

The key issue with respect to commercial use of nanotubes for most applications comes in placing wafers on silicon or silicon on insulators, which are typical substrates for making circuits or devices. To the extent that alignment or specific orientation is required, applicability has been limited. Two general methodologies exist for proceeding toward manufacturability of devices: specific growth using prepatterned catalysts (Figure 13–4), or patterning of nonoriented fabrics using application of solutions. In the former case catalysts are placed in specific regions of a substrate onto which nanotubes grow in either a vertical or a horizontal direction.

Vertically oriented growth has been demonstrated by several groups and is especially valuable for field emitting devices. Many potential CNT device approaches using CVD growth of SWNTs suffer from manufacturability and

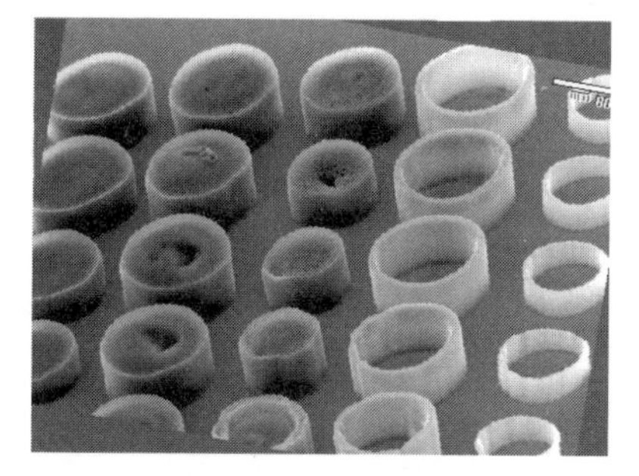

Figure 13–4 Patterned growth of multiwalled carbon nanotubes by chemical vapor deposition.

scalability issues, primarily because typical CVD temperatures are >800°C for SWNTs. At such high temperatures, other steps in a fabrication process can be adversely affected, causing the yields of working devices to be lower. Indeed this a serious limitation if the underlying substrate or control electronics cannot withstand such temperatures. Some reports have presented data on lower-temperature methods of growing SWNTs using techniques such as PECVD. These techniques are still in their infancy but could represent a reasonable pathway for certain types of devices or applications.

Horizontal fabrics have been applied to substrates whose thickness can be controlled by solution concentration and application procedure. Such methods have conquered the issues of purification of nanotube solutions, distribution of networks of nanotubes of uniform density over large substrates, and patterning of traces of conductive nanotubes into shapes that can be further integrated into complicated process flows.

SWNTs can be obtained in bulk from various suppliers, which have exploited advanced CVD techniques to grow SWNTs in large scale of excellent quality. These nanotubes can then be purified to remove metallic contaminants (for example, Group IA and IIA elements) used in the growth process and carbonaceous species that serve as a potential source of contamination. The processed SWNTs are solubilized and can then be applied to device substrates for further processing.

Potential Applications

A variety of potential applications exist for carbon nanotubes. MWNTs have been reported for use primarily as a composite for batteries and as field emitters for television monitors.[6] As the price for these materials continues to drop, other potential applications, especially as additives in composites, are likely.

The molecular nature of carbon nanotube fabrics allows various CNT physical properties, including electromagnetic, mechanical, chemical, and optical behaviors, to be exploited to create integrated electronic devices (including nonvolatile memory devices); chemical, biological, and radiation sensors; passive low-resistance, low-capacitance conformal interconnects; and electromagnetic field emission devices, scaffolds for cell growth, antioxidants, and near infrared imaging tags for biological samples and cells, to name a few. Table 13–1 summarizes a number of proposed applications that could use SWNT fabrics as an enabling component.[7]

One existing application for which SWNTs are particularly useful is in the arena of electronics, specifically to create nonvolatile memory.[8] In the case of nonvolatile memory applications, significant progress has been made in

Table 13–1 Sample applications of single-walled carbon nanotube fabrics.

Semiconductors	*Life Sciences*	*Instrumentation*
Nonvolatile and volatile memory	Membranes	Mechanical relays and switches
Programmable logic devices	Chemical absorption	Thermal sensors and actuators
Global and local interconnects	Valves	Acceleration sensors
Inductors	Nanomixers	Bolometers
Radio frequency (RF) components	Heat exchangers	Gyroscopes
Micromirrors	Nanochannels	Field emission tips displays
Chip-to-chip interconnects	Reaction chambers	Acoustic/Pressure sensors
Transistors	Nanofluidics devices	Radiation detectors

using fabrics, or assemblages of SWNTs, as electrical traces within integrated circuits.[9] These fabrics retain their molecular-level properties while eliminating the need for nanoscale physical control. These monolayers are created by room-temperature spin-coating of a solution of SWNTs in a semiconductor-grade solvent. After evaporation of the solvent, the resulting monolayer fabric is lithographically patterned and etched in an oxygen plasma. The process of spin-coating SWNT solutions can be used to produce monolayer fabric of very steep aspect ratios, which allows coverage of fabrics over sharp edges or tall structures. Such diversity in the coating of nanotube fabrics on three-dimensional structures has the potential to support traditional horizontal integration schemes in electronics as well as a novel set of devices oriented in a vertical fashion that could lead to significantly more dense electronics.

The combination of patterned nanotube fabrics and the use of fabricated sacrificial layers around the fabric allows the formation of an open cavity, in which a suspended patch of fabric can be mechanically drawn into electrical contact with an electrode. These devices, called molecular microswitches (MMSs), can be used as storage elements for memory or logic applications.

The SWNT fabric approach uses a 1–2-nm thick patterned SWNT fabric and can be interconnected monolithically with additional standard semiconductor (for example, CMOS) circuitry above or below, providing buffering and drive circuitry to address and control the MMS devices. When this technique is used, the limit to the scalability of the hybrid nanotube/CMOS system depends only on the available photolithographic node. Hence, the CMOS fabrication, and not the inclusion of nanotubes, remains the limiting factor in scaling. Indeed, the ultimate physical limit to the integration density of this approach scales down to two individual metallic nanotubes and their electromechanical interaction.

REFERENCES

1. S. Iijima, *Nature* 354, (1991): 56.
2. C. Journet et al., *Nature* 388 (1997): 756; A. Thess et al., *Science* 273 (1996): 483; and A. M. Cassell, J. A. Raymakers, J. Kong, and H. Dai, *J. Phys. Chem. B* 103 (1999): 6484.
3. T. W. Ebbesen et al., *Nature* 382 (1996): 54; M. M. J. Treacy, T. W. Ebbesen, and J. M. Gibson, *Nature* 381 (1996): 678; and J. W. Mintmire, B. I. Dunlap, and C. T. White, *Phys Rev Lett* 68 (1992): 631.
4. http://www.lenntech.com/Periodic-chart-elements/melting-point.htm.
5. D. Pantarotto et al., *Chem Biol.* 10 (2003): 961.
6. Y. Chen et al., *Appl. Phys. Lett.* 78 (2001): 2128; and D. Qian et al., *Appl. Phys. Lett.* 76 (2000): 2868.
7. R. H. Baughman, A. A. Zakhidov, and W. A. de Heer, *Science* 297 (2002): 787.
8. T. Rueckes et al., Science 298 (2000): 94.
9. D. K. Brock et al., Proc. 2005 IEEE Aero. Conf. (2005).

NANOWIRES

Zhong Lin Wang

Semiconductor nanowires (NWs) are wires only a few nanometers in size that do not occur naturally. They represent an important broad class of nanometer-scale wire structures, which can be rationally and predictably synthesized in single-crystal form with all their key parameters controlled during growth: chemical composition, diameter, length, doping, and so on. Semiconductor NWs thus represent one of best-defined and best-controlled classes of nanoscale building blocks, which correspondingly have enabled a wide range of devices and integration strategies (Figure 13–5). For example, semiconductor

Figure 13–5 Aligned semiconducting ZnO nanowire arrays. These wires are grown uniformly from a solid substrate, and their location and density can be defined by the deposited catalyst of gold.

NWs have been assembled into nanometer-scale field effect transistors (FETs), p-n diodes, light-emitting diodes (LEDs), bipolar junction transistors, complementary inverters, complex logic gates, and even computational circuits that have been used to carry out basic digital calculations. It is possible to combine distinct NW building blocks in ways not possible in conventional electronics. Leveraging the knowledge base of chemical modifications of inorganic surfaces can produce semiconductor NW devices that achieve new functions and produce novel device concepts.

Semiconducting oxide nanobelts (NBs) are another unique group of quasi-one-dimensional nanomaterials with well-defined side surfaces, which have been systematically studied for a wide range of materials having distinct chemical compositions and crystallographic structures. Beltlike, nanobelts (also called nanoribbons) have been synthesized for semiconducting oxides of zinc (Figure 13–6), tin, indium, cadmium, and gallium by simply evaporating the desired commercial metal oxide powders at high temperatures. These oxide nanobelts are pure, structurally uniform, single-crystalline, and largely free of imperfections; they have a rectangularlike cross section with controllable dimension. Field effect transistors and ultrasensitive nanosize gas sensors, nanoresonators, and nanocantilevers have also been fabricated based on individual nanobelts. Thermal transport along the nanobelt has also been measured. Nanobelts, nanosprings, and nanorings that exhibit piezoelectric properties have been synthesized that could eventually be used to make nanoscale traducers, actuators, and sensors.

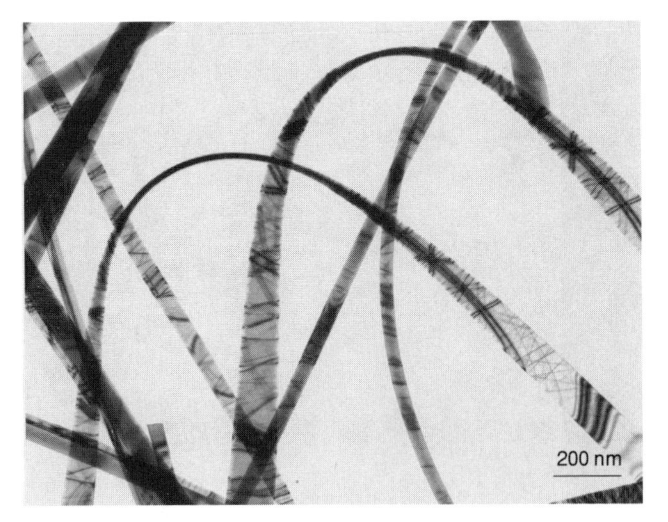

Figure 13–6 Nanobelt of ZnO synthesized by a vapor-solid process. The nanobelt has a well-defined side surface, and it usually has a rectangular cross section.

Applications

Biological Sensors Based on Individual Nanowires

Because many biological molecules, such as proteins and DNA, are charged under physiological conditions, it is also possible to detect the binding of these macromolecules to the surface of nanowires. Protein detection by SiNW nanosensors has been demonstrated for the well-characterized ligand-receptor binding of biotin-streptavidin (Figure 13–7) using biotin-modified NW surfaces.

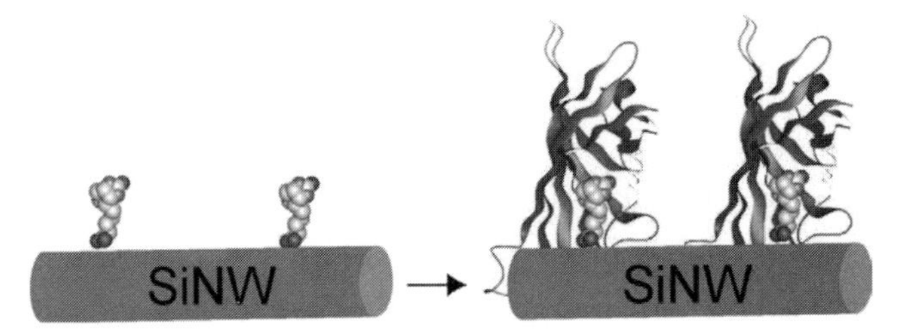

Figure 13–7 Detection of protein binding. The schematic illustrates a biotin-modified SiNW (left) and subsequent binding of streptavidin to the SiNW surface (right). The SiNW and streptavidin are drawn approximately to scale.

Crossed p-n Junction Light-Emitting Diodes

In direct band gap semiconductors like InP, the p-n junction also forms the basis for critical optoelectronic devices, including LEDs and lasers (see Figure 13–8). To assess whether our crossed NW devices might behave similarly, the photoluminescence (PL) and electroluminescence (EL) from crossed NW p-n junctions can provide a means for controlling the color of the LEDs in a well-defined way. The ability to tune color with size in these nanoLEDs might be especially useful in future nanophotonic applications.

Integrated Nanowire Logic Devices

The controlled high-yield assembly of crossed NW p-n diodes and crossed-nanowire field effect transmitters with attractive device characteristics, such as high gain, enable the bottom-up approach to be used for assembly of more complex and functional electronic circuits such as logic gates. Logic gates are critical blocks of the hardware in current computing systems that produce a logic-1 and logic-0 output when the input logic requirements are satisfied.

Diodes and transistors represent two basic device elements in logic gates. Transistors are more typically used in current computing systems because they can exhibit voltage gain. Diodes do not usually exhibit voltage gain, although they may also be desirable in some cases. For example, the architecture and constraints on the assembly of nanoelectronics might be simplified using diodes because they are two-terminal rather than three-terminal transistors. In addition, by combining the diodes and transistors in logic circuits,

Figure 13–8 Three-dimensional plot of light intensity of the electroluminescence from a crossed NW LED.

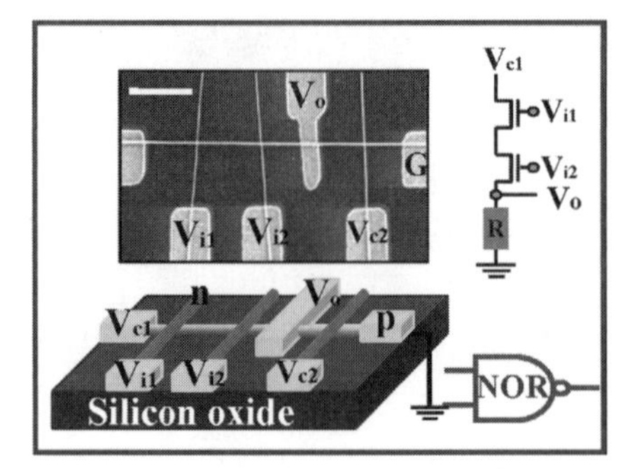

Figure 13–9 Logic NOR gate. Schematic of logic NOR gate constructed from a 1x3 crossed NW junction array. The insets show a sample scanning electron microscope image (bar is 1 m) and symbolic electronic circuit.

it is possible to achieve high voltage gain while simultaneously maintaining a simplified device architecture. The flexibility of these NW device elements for both diode- and FET-based logic has been demonstrated in the creation of AND, OR, NOR gates that have been used to implement simple basic computation (Figure 13–9).

Polar-Surface Dominated Novel Nanostructures

For applications in nanotechnology, ZnO has three key advantages. First, it is a semiconductor that has a direct wide-band gap and significant excitation-binding energy. It is an important functional oxide, exhibiting near-UV emission and transparent conductivity at room temperature and higher. It is also piezoelectric because of its noncentral symmetry. This is a key phenomenon in building electromechanical coupled sensors and transducers at nanoscale. The piezoelectric coefficient of a polar nanobelt is about three times that of bulk material, and this makes it a candidate for nanoscale electromechanical coupling devices. Finally, ZnO may be biosafe and biocompatible, so it can be used for biomedical applications without coating. With these three unique characteristics, ZnO could be one of the most important nanomaterials in future research and applications. The diversity of nanostructures presented here for ZnO should open many fields of research in nanotechnology.

The synthesis of various nanostructures is based on a solid-state thermal sublimation process, in which a pile of source materials, typically the powder form of oxides, is placed at the center of a tube furnace. The source materials are sublimated by raising the temperature of the furnace; a redeposit ion of the vapor phase at a lower temperature zone produces novel nanostructures. By controlling the growth kinetics, the local growth temperature, and the

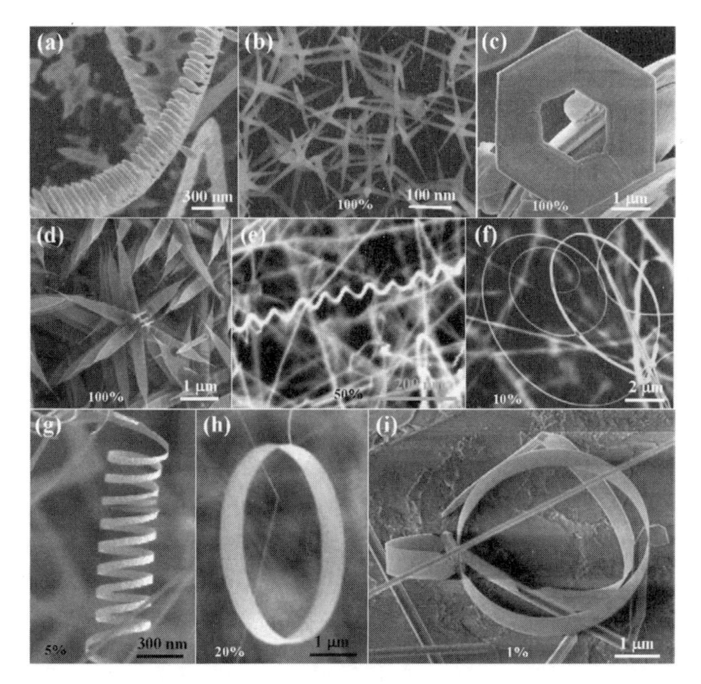

Figure 13–10 A collection of polar-surface induced/dominated nanostructures of ZnO, synthesized under controlled conditions by thermal evaporation of solid powders unless specified otherwise: (a) nanocombs induced by asymmetric growth on the Zn-(0001) surface; (b) tetraleg structure due to catalytically active Zn-(0001) surfaces; (c) hexagonal disks or rings synthesized by solution-based chemical synthesis; (d) nanopropellers created by fast growth; (e) deformation-free nanohelixes as a result of block-by-block self-assembly; (f) spiral of a nanobelt with increased thickness along the length; (g) nanosprings; (h) single-crystal seamless nanoring formed by loop-by-loop coiling of a polar nanobelt; (i) a nanoarchitecture composed of a nanorod, nanobow, and nanoring. The percentage in each figure indicates the purity of the as-synthesized sample for the specific nanostructure in a specific local temperature region.

chemical composition of the source materials, researchers have synthesized a wide range of polar-surface dominated nanostructures of ZnO under well-controlled conditions at high yield (Figure 13–10).

Zinc oxide (ZnO) is a material that has key applications in catalysts, sensors, lasering, piezoelectric transducers, transparent conductors, and surface acoustic wave devices. This chapter focuses on the formation of various nanostructures of ZnO, showing their structures, growth mechanisms, and potential applications in optoelectronics, sensors, transducers, and biomedical science. For the nonpolar surfaces of ZnO, various forms of nanostructures have been synthesized for ZnO (Figure 13–11). These structures were realized by controlling growth kinetics.

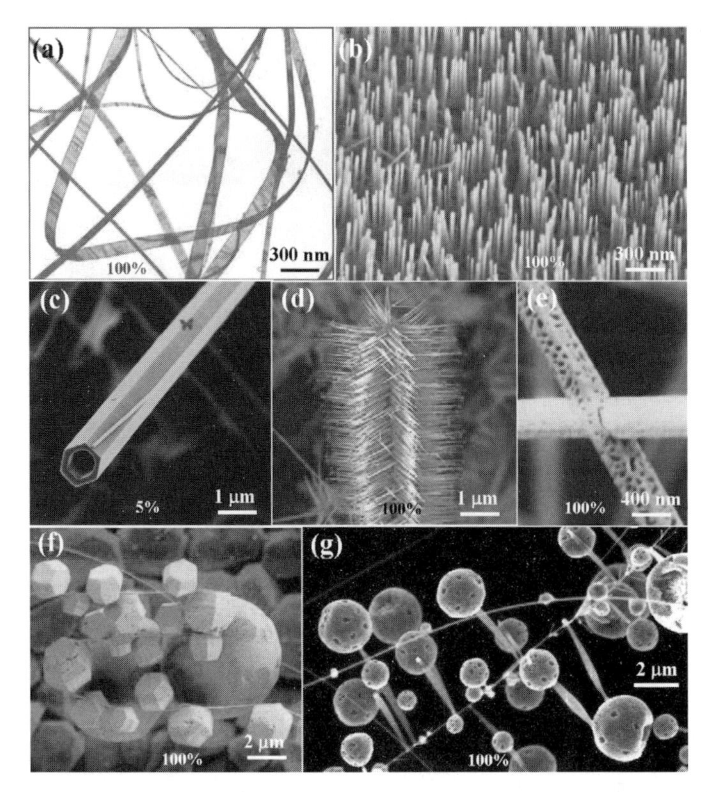

Figure 13–11 A collection of nonpolar-surface dominated nanostructures of ZnO, synthesized under controlled conditions by thermal evaporation of solid powders by controlling source materials, growth temperature, temperature gradient, and substrate. (a) Nanobelt; (b) aligned nanowire arrays; (c) nanotubes; (d) array of propellers; (e) mesoporous nanowires; (f) cages and shell structures; and (g) hierarchical shell and propeller structure. The percentage in each figure indicates the purity of the as-synthesized sample for the specific nanostructure in a specific local temperature region.

SOFT NANOTECHNOLOGY

Fiona Case

Many soft or fluid consumer products—such as foods, paint, detergents, personal care products, and cosmetics—contain nanometer- to micrometer-scale structures. These structures are formed by the spontaneous self-assembly of natural or synthetic surfactants or block copolymers. In many cases, to create the desired structure and performance, complex mixtures of different surfactants and polymers are required.

Figure 13–12 shows an example of a simple, nonionic surfactant. Figure 13–13 illustrates some of the soft nanoscale structures that form spontaneously in surfactant or block copolymer solution. The particular structure that is formed depends on the relative sizes of the head and tail and their chemical character. For example, surfactants that have ionic (charged) head groups often form spherical micelles, whereas non-ionic (uncharged) surfactants are more likely to form rodlike structures, and surfactants that have more than

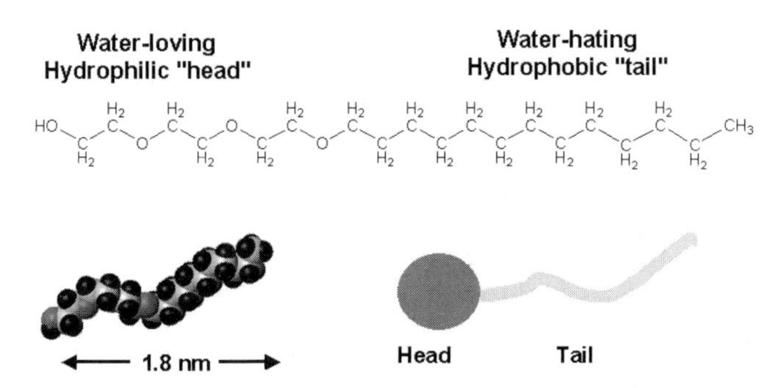

Figure 13–12 A simple example of a surfactant, in this case a non-ionic ethoxylate surfactant, shown as the chemical structure (top), as a space-filling molecular model (bottom left), and as a cartoon representation (bottom right).

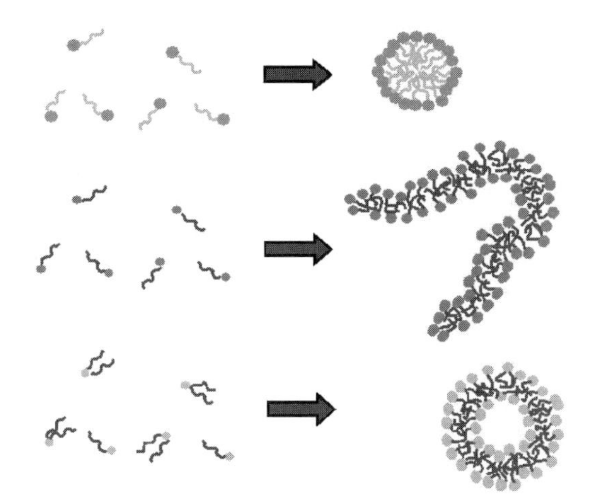

Figure 13–13 Cartoon representations of different surfactants forming a spherical micelle (top), a rodlike micelle (middle), and a vesicle (bottom) in water. A slice through each of the nanostructures is shown; in the complete structures, the hydrophilic head groups surround the hydrophobic tails, shielding them from unfavorable interactions with water.

one tail are more likely to form vesicles or lamella (almost flat sheets). The soft nanoscale structure can also be influenced by the order of addition of the different ingredients.

These soft nanostructures do work on demand. When you add dish liquid to hot water in a dishpan full of dirty pots and pans, the nanoscale structure of the surfactant micelles change as they migrate to the oily surfaces, encapsulate the grease, and carry it into solution. When you paint a wall with a good-quality emulsion-based paint, the brush moves easily as you spread the paint on the wall, but as soon as you finish brushing an area, the structure changes and the viscosity of the paint increases to reduce the occurrence of drips. Soft nanotechnology is also at work when you mix up a cake or even enjoy the taste and texture of chocolate, ice cream, or yogurt.

Designing these complex materials is difficult, and characterizing them is challenging. Until recently, most soft nanostructured products were created by experienced formulators, using empirical rules that were passed down through generations of product developers. Although ideas of self-assembly were invoked to explain behavior, it was almost impossible to predict what structure would be created from a particular mixture of ingredients. And because it was extremely difficult to characterize products that were known to be successful, it was hard to know what structure was desirable for a particular application.

Today, new methods of characterization are providing insight into soft nanostructured materials—for example, dynamic light scattering, NMR diffusion experiments, X-ray and neutron scattering, and electron microscopy. New robotics techniques are allowing developers to explore many more formulas than was possible when each candidate formula had to be created by hand. New theories and computer simulation methods are enabling researchers to construct realistic models of soft nanostructured materials. Taken together, these developments have the potential to revolutionize consumer product development.

Other members of the nanotechnology community are also becoming interested in self-assembly and nanostructured fluids, which indeed have the potential to provide robust and inexpensive strategies for creating nanoscale materials. For example, IBM recently highlighted the potential for block-copolymer self-assembly to create nanometer-size structures for electronics applications. Several groups are exploring strategies for creating hard nanoparticles using soft self-assembled structures as templates, and self-assembled block copolymer-stabilized emulsions and vesicles are being proposed as drug- (and nutriceutical-) delivery devices. Moreover, various soft nanotech strategies are being applied to stabilize colloidal systems (nanoparticle solutions and gels).

Chapter 14

Nanotechnology-Enabled Sensors: Possibilities, Realities, and Diverse Applications

David J. Nagel and Sharon Smith

Whether you are a producer or a user of sensors, your business will likely be impacted by current and future developments in nanotechnology. Emerging nanotechnologies offer an unprecedented promise for sensors: smaller size and weight, lower power requirements, more sensitivity, and more specificity, to name a few.

This chapter surveys both the promise and the limitations of the new sensors that are enabled by technologies on the scale of nanometers (one-thousandth of a micrometer). Nanosensors and nano-enabled sensors have applications in a wide variety of industries, including transportation, communications, building and facilities, medicine, safety, defense, and national security. Proposed applications of nanoscale sensors are far reaching and include using nanowire nanosensors to detect chemicals and biologics, putting nanosensors in blood cells to detect early radiation damage in astronauts, and using nanoshells to detect and destroy tumors.[1] Many such applications are being developed by start-up companies, which seek to build businesses that exploit particular nanoscale effects or particular applications of nanotechnology.

The expected impact of nanotechnology on sensors can be understood by noting that most chemical and biological sensors, as well as many physical sensors, depend on interactions that occur on the scale of atoms and molecules. Nanotechnology refers to an expanding science, and a very promising technology, that enables the creation of functional materials, devices, and systems through control of matter at the scale of atoms and molecules, and the exploitation of novel properties and phenomena at the same scale.[2]

Nanotechnologies can extend the long-established trend toward smaller, faster, cheaper materials and devices. The miniaturization of macro techniques led to the now established field of microtechnologies. For example, electronic, optical, and mechanical microtechnologies have had a dramatic impact on the sensor industry in recent decades. Improved, new, and smart sensors are among the many beneficial effects of integrated circuits, fiber, and other microoptics and MEMS (microelectromechanical systems). Now, developments in nanotechnology will drive the devices into even smaller dimensions, with significant potential advantages.

As the trend continues toward the use of smaller building blocks of atoms and molecules, there will also tend to be convergence of technology disciplines—for example, the merging of nanotechnology, biotechnology, and information technology. This overlap and the resulting synergy should contribute to the growing strength of each of these major technologies.

POSSIBILITIES

The global excitement over nanotechnology has resulted from several discoveries late in the twentieth century. Primary among them is the ability to manipulate individual atoms in a controlled fashion, a sort of atomic bricklaying made possible by techniques such as scanning probe microscopy (SPM). The ability to produce significant amounts of nanoparticles, such as silver and gold nanoparticles, has also fed the excitement. And the underlying fact that materials and devices on the scale of atoms and molecules have new and useful properties—due in part to quantum and surface effects—can also be cited. Nanoscale quantum dots with remarkable optical properties are a prime example.

Another major contributor to the interest in nanotechnology was the discovery of carbon nanotubes (CNTs), which are extremely narrow, hollow cylinders made up of carbon atoms. With the possibility of single- and multiple-walled varieties, each with a myriad of potential applications, carbon nanotubes offer many exciting prospects. The single-walled versions can have various geometries, as shown in Figure 14–1. Depending upon the exact orientation of the carbon atoms, the nanotubes can exhibit either conducting (metallic) or semiconducting properties. This fact, plus the ability to grow CNTs at specific locations and to manipulate them after growth, leads to optimism that CNTs will be important for electronics and sensors, either alone or integrated with each other.

For these and other reasons, the funding for nanotechnology has increased more than a factor of 5 in the period 1997 to 2003, and it is still

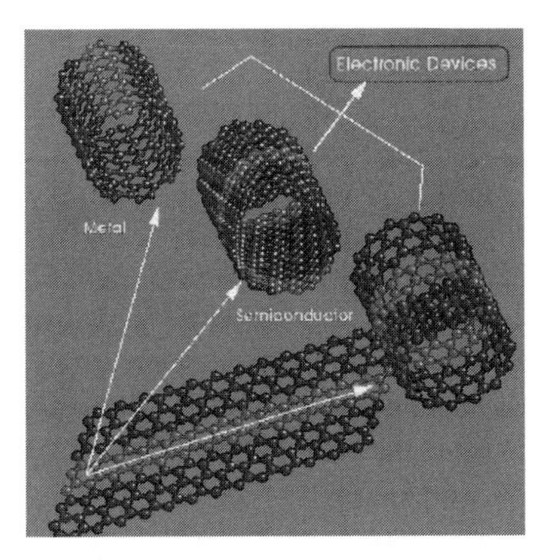

Figure 14–1 Carbon nanotubes can exist in a variety of forms, which can be either metallic or semiconducting in nature. (Courtesy of NASA Ames Research Center, Moffett Field, CA.)

increasing.[3] With the major increase in funding and the resulting scientific breakthroughs, there is great potential for improvements in existing sensors and for the development of really new sensors. One market study of nano-enabled sensors predicts sales near $0.6 billion by 2009.[4] Another is much more aggressive, saying that the nanotechnology sensor market will generate global revenues of $2.7 billion by 2008.[5] Whatever the actual outcome, it seems clear that the sales of nano-enabled sensors will grow significantly, and the growth is largely a result of the research investment in nanoscience and nanotechnology.

Relentless Integration

Historically, there has been a relatively clean separation between the materials, device, and systems levels. However, in recent decades, integration has been a major theme of technology. First, transistors were made into integrated circuits. Then such microelectronics were integrated with microoptics and micromechanics. The microtechnology devices were packaged individually and mounted on printed circuit boards. The recent use of flip chips—where the chip is the package—and the placement of passive components *within* printed circuit boards are blurring the classical distinction between devices and systems. The integration of nanomaterials into MEMS devices is now a major research topic. Nanotechnology makes it possible to integrate the different levels to the point that the (very smart) material essentially *is* the device and possibly also the system. Larry Bock, founder of Nanosys, recently noted, "Nanotech takes the complexity out of the system and puts it in the material."[6]

Now the vision of sensing the interaction of a small number of molecules, processing the data, transmitting the information using a small number of electrons, and then storing information in nanometer-scale structures can be seriously contemplated. Fluorescence and other means of single-molecule detection are being developed. Data storage systems, which use proximal probes to make and read indentations in polymers, are being developed by IBM and others. They promise read and write densities near one trillion bits per square inch, far in excess of current magnetic storage densities.[7]

The integration of such nanoscale capabilities, once they are adequately developed, is challenging, but it would result in capable smart sensors that are small, use low power, and can be used cheaply in large numbers. These advances will enable, for example, more in-situ sensing of structural materials, more sensor redundancy in systems, and increased use of sensors in size- and weight-constrained systems, such as satellites and space platforms.

The production of nano-enabled sensors depends on the availability of nanomaterials and processes to grow them in place or else the practice of placing previously grown materials in the right positions in the desired amounts and geometries. Hence, there is a three-dimensional space that interrelates nanomaterials, processes for making or handling them, and specific sensors. This is sketched in Figure 14–2. What materials are used, how they are employed, and why a particular sensor is being produced all come into consideration. A given material might be made or handled by a variety of processes and made into different sensors. A particular process might be germane to various materials and sensors.

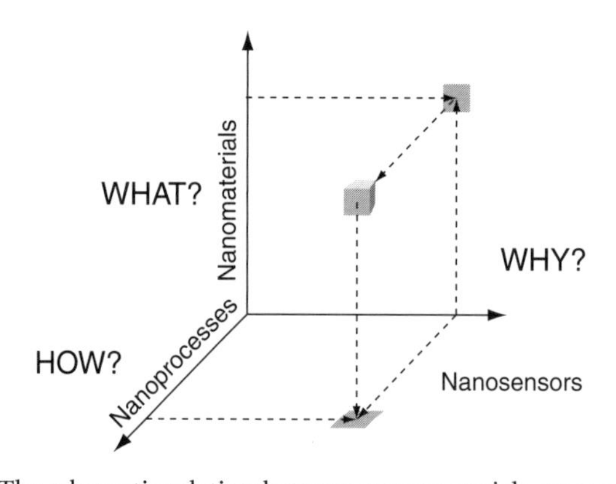

Figure 14–2 The schematic relation between nanomaterials, processes for making or emplacing them, and the sensors that are thus enabled.

Advances in Processing

Recent progress in "top-down" manufacturing processes has helped enable microtechnologies as well as nanotechnologies. Leading-edge integrated circuits are made by top-down processes in which lithography, etching, and deposition techniques are used to sculpt a substrate, such as silicon, and then build up structures on it. Using these approaches, conventional microelectronics has approached the nanometer scale as the line widths in chips surpass the 100-nm level and are continuing to shrink. MEMS devices are constructed in a similar top-down process. As these processes continue to shrink to smaller and smaller dimensions, they can be used for making a variety of nanotechnology components, much as a large lathe can be used to make small parts in a machine shop.

In the nanotechnology arena, such processes are being joined by a collection of "bottom-up" methods in which individual atoms and molecules are used to construct useful structures. Under the right conditions, the atoms, molecules, and larger units can self-assemble.[8] Alternatively, directed assembly techniques can be employed.[9]

The control of both self-assembly and directed assembly presents major technical challenges. Processes for getting nanomaterials to grow in the right places in the desired geometries require development. Overall, the combination of nanoscale top-down and bottom-up processes gives the designer of materials and devices a wide variety of old and new tools. It is also possible to combine micro- and nanotechnologies to enable the development of new sensor systems.

Diverse Nanomaterials

Nanomaterials and structures offer other possibilities for sensors. There are two often separable functions within many sensors, especially those for chemicals and biological materials. They are *recognition* of the molecule of interest, and *transduction* of that recognition event into a useful signal. Nanotechnology offers new possibilities in both of these fundamental arenas in a manner that offers benefits far beyond the advantages offered by MEMS and other microsensors.

Most of the nanomaterials that have been produced and used to date contain only one kind of nanostructure and chemistry. There are a few exceptions, but the situation for nanomaterials is now similar to that for macromaterials before composites, such as fiberglass, were formulated. There is no reason that composites of two or more nanomaterials should not be made, tested, and exploited.

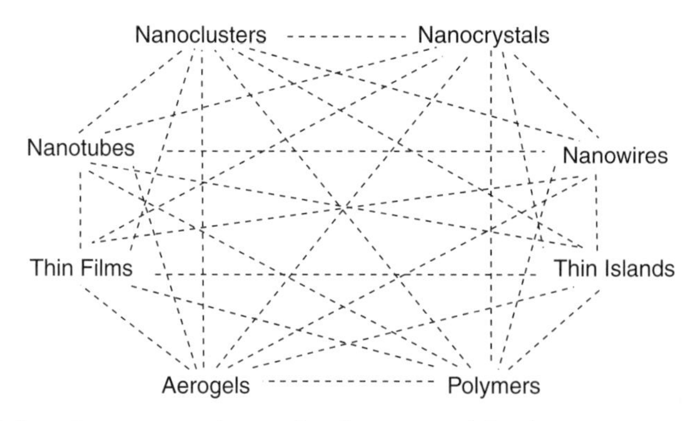

Figure 14–3 Binary composites made of nanomaterials taken two at a time. The top row contains zero-dimensional (0-D) materials, the second row 1-D materials, the third row 2-D materials, and the bottom row 3-D materials. The binaries represent 28 possible composite materials.

As with large-scale materials, the combinatorics of making composite materials results in a very large number of potential new materials. This is illustrated in Figure 14–3 for composite materials made of two nanomaterials. Making nanocomposites from different kinds of nanomaterials, which can have a wide range of chemistries, presents many processing challenges. It is expected that the production and use of composite nanomaterials for sensors and other applications will increase steadily in the coming years.

New Tools

Another aspect of the development of nanotechnology-based sensors deserves emphasis. Because of experimental tools now available—notably synchrotron X-radiation and nuclear magnetic resonance—the atomic structures of many complex molecules are now known. But knowledge of the structure on the nanometer scale is not enough. It is the interactions of atoms and molecules that are important in the recognition, and sometimes the transduction, stage of sensing. The availability of powerful computers and algorithms for the simulations of interactions on the nanoscale makes it possible to design nanotech-based sensors computationally, and not just experimentally. Molecular dynamics codes and calculations are already a fundamental tool in nanoscience, and they can be applied to the development of nanotechnology-based sensors.[10]

REALITIES

Although the excitement about nanotechnology and its prospective uses is generally well founded, there are many inescapable requirements that must be considered for the successful development and integration of nanotechnology sensors. These requirements include those imposed by physics, chemistry, biology, engineering, and commerce.

For example, as nanotechnologies are integrated into macro-sized systems, it will be necessary to provide for and control the flow of matter, energy, and information between the nanometer-scale and macroscale systems. Calibration of nano-enabled sensors can be problematic if very small amounts of analyte and reagent materials are involved. It must be remembered that the size of most sensor systems is determined by the scale of the computer, memory, and radio chips, and (especially!) by the batteries and antennas. This is true for both micro- and nanoscale sensors.

Intensified Design Problems

Many of the design considerations for nanoscale sensors are similar to those of microsensors—notably, interface requirements, heat dissipation, interferants, and noise (both electrical and mechanical). Each interface in a microsystem is subject to unwanted transmission of electrical, mechanical, thermal, and possibly chemical, acoustical, and optical fluxes. Dealing with unwanted molecules and signals in very small systems often requires ancillary equipment and the use of low temperatures to reduce noise.

Flow control is especially critical in chemical and biological sensors into which gaseous or liquid analytes must be brought and from which they are expelled. The very sensitive, tailored surfaces in molecular sensors, chemical as well as biological, are subject to degradation by interfering substances, heat, and cold. One of the attractions of nanotechnology is the chance to make hundreds of sensors in a small space so that sensors that have degraded can be ignored in favor of fresh sensors to prolong the useful lifetime of a system containing them.

The Risks of Commercialization

The path from research to engineering to products to revenues to profits to sustained commercial operations is difficult for technologies of any scale. It is particularly challenging for new nanotechnologies, and not only because of the common reluctance to specify new technologies for high-value systems.

The reality is that most nanoscale materials are currently difficult to produce in large volumes, so unit prices are high and markets limited. Costs will decrease over time, but it may be difficult for small companies to make enough profit soon enough. A recent review has focused on the commercialization of nanosensors.[11]

DIVERSE APPLICATIONS

Although there are few sensors today based on pure nanoscience and it is early in the development of nano-enabled sensors, some of the possible devices and applications are already clear. Examples of nanotech and nano-enabled physical, chemical, and biological sensors are given in this section. Although sensors for physical quantities were the focus of early development efforts, nanotechnology will contribute heavily to realizing the potential of chemical and biosensors for safety, medical, and other purposes. Vo-Dinh, Cullum, and Stokes recently provided an overview of nanosensors and biochips for the detection of biomolecules.[12] It is anticipated that many major industries will benefit from nanotechnology-enabled sensors.

Physical Sensors

An early example of a nanotech physical sensor is given in Figure 14–4. Researchers at the Georgia Institute of Technology demonstrated the world's smallest balance by taking advantage of the unique electrical and mechanical properties of carbon nanotubes.[13] They mounted a single particle on the end of a carbon nanotube and applied an electric charge to it. The carbon nanotube acted much like a strong, flexible spring. The nanotube oscillated without breaking, and the mass of the particle was calculated from changes in the resonance vibrational frequency with and without the particle. This approach may lead to a technique for the weight measurement of individual biomolecules.

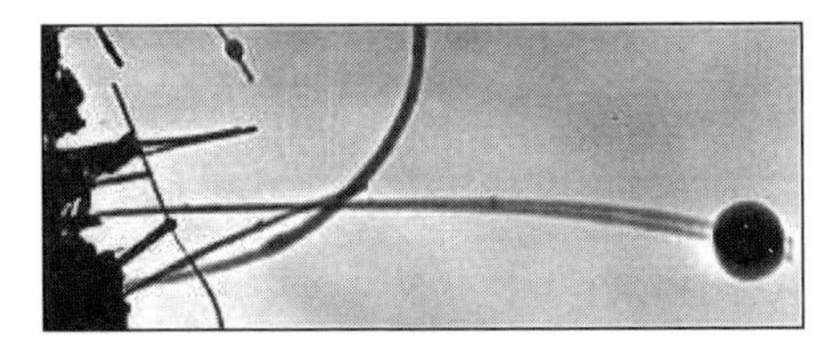

Figure 14–4 The mass of a carbon sphere shifts the resonance frequency of the carbon nanotube to which it is attached. (Courtesy of Walter de Heer, Georgia Institute of Technology, Atlanta, GA.)

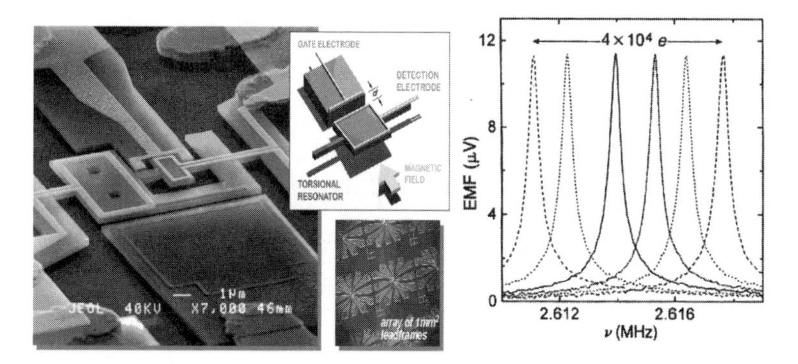

Figure 14–5 A nanometer-scale mechanical electrometer consists of a torsional mechanical resonator, a detection electrode, and a gate electrode used to couple charge to the mechanical element. (Reprinted with copyright permission from Nature Publishing Group.)

Cleland and Roukes at the California Institute of Technology reported the fabrication and characterization of a submicron, mechanical electrometer.[14] Shown in Figure 14–5, this device, made possible by modern nanofabrication technology, has demonstrated charge sensitivity less than a single electron charge per unit bandwidth (~0.1 electrons/sqrt (Hz) at 2.61MHz). This sensitivity exceeds that of state-of-the-art semiconductor devices.

Chemical Sensors

A variety of nanotube-based gas sensors have been described in the past few years. Modi et al. have used carbon nanotubes to develop a miniaturized gas ionization detector.[15] This system could be used as a detector for gas chromatography. Titania nanotube hydrogen sensors have been employed in a wireless sensor network for detecting hydrogen concentrations in the atmosphere.[16] And Kong et al. have developed a chemical sensor for gaseous molecules, such as NO_2 and NH_3, based on nanotube molecular wires.[17] In a recent study, Parikh et al. used office equipment to print carbon nanotubes on flexible plastic to make sensors for organic vapors.[18]

Sensors based on nanocantilevers also are possible. Datskos and Thundat have fabricated nanocantilevers using a focused ion beam (FIB) technique.[19] They developed an electron transfer transduction approach to measure the motion of the nanocantilever. An array of such sensors is shown in Figure 14–6. They may provide the sensitivity to detect single chemical and biological molecules. Structurally modified semiconducting nanobelts of ZnO have also been demonstrated to be applicable to nanocantilever sensor applications.[20]

Figure 14–6 This array of chemical sensors incorporates MEMS cantilevers and the electronics for signal processing. (Courtesy of Thomas G. Thundat, Oak Ridge National Laboratory, Oak Ridge, TN.)

Biosensors

Nanotechnology will also enable the very selective, sensitive detection of a broad range of biomolecules. It is now possible to create cylindrical rods made up of metal sections that are between 50nm and 5 μm long. This is done by the sequential electrochemical reduction of the metal ions onto an alumina template. These particles, called Nanobarcodes, can be coated with analyte-specific entities, such as antibodies, for selective detection of complex molecules, as illustrated in Figure 14–7. DNA detection with these nanoscale coded particles has been demonstrated.

Researchers at NASA Ames Research Center have taken a different approach to DNA detection.[22] Carbon nanotubes in the range of 30–50nm in diameter are mounted vertically on a silicon chip, with millions of nanotubes covering the surface of the chip, as shown in Figure 14–8. Probe DNA mole-

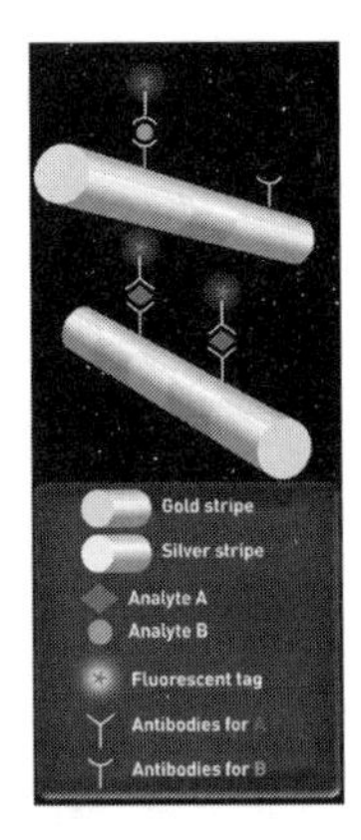

Figure 14–7 Biomolecules can be detected optically when they are attached to antibodies affixed to particles called Nanobarcodes.

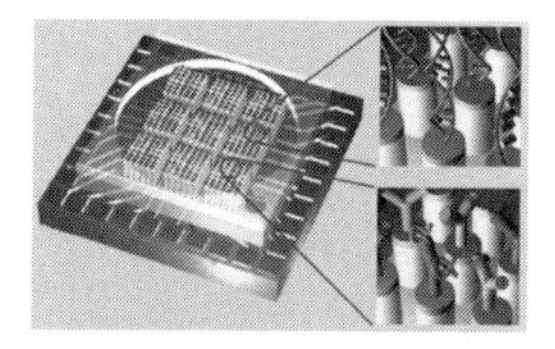

Figure 14–8 Vertical carbon nanotubes are embedded in a silicon chip. DNA molecules attached at the ends of the nanotubes detect specific types of DNA or other substances. (Courtesy of NASA Ames Research Center, Moffett Field, CA.)

cules are attached to the ends of the nanotubes. When the chip is placed in a liquid containing DNA molecules, the substrate DNA attaches to the target DNA, increasing the measured electrical conductivity. This approach, expected to reach the sensitivity of fluorescence-based detection systems, may find application as a portable sensor.

Deployable Sensors

We are starting to see nanoscale materials and devices being integrated into real-world systems. An example is SnifferSTAR, a lightweight and portable chemical detection system, shown in Figure 14–9.[23] This unique system combines a nanomaterial for sample collection and concentration with a MEMS-based chemical lab-on-a-chip detector. SnifferSTAR has defense and homeland security applications and is ideal for deployment on unmanned systems such as micro-UAVs (unmanned aerial vehicles).

Other areas expected to benefit from sensors based on nanotechnologies include the following: transportation (land, sea, air, and space); communications (wired and wireless, optical and radio frequency); buildings and facilities (homes, offices, factories, and other structures); monitoring of humans (especially health and medical monitoring of aging baby boomers); and robotics of all types. The integration of nano-enabled sensors into commercial and military products will be fueled by many new companies making nanomaterials and some making sensors based on them.

Although much progress is being made, many challenges still must be met before we will see the full impact of nanotechnology in the sensor industry. They include reducing the cost of materials and devices, improving reliability, and the packaging of the devices into useful products. Such challenges notwithstanding, the future of nano-enabled sensors is bright for several reasons. It may be possible to put sensors into very small volumes, including individual cells. Nanosensors may have a significant scientific impact by measuring molecular interactions and kinetics on very small scales at high speeds.

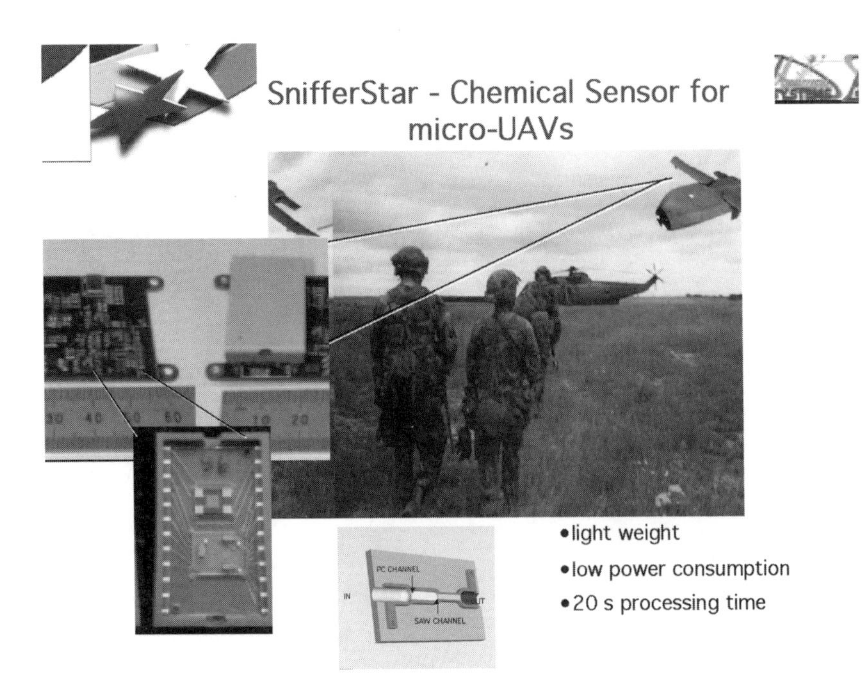

Figure 14–9 A nano-enabled chemical sensor has been integrated into a micro-UAV. (Courtesy of Sandia National Laboratories, Albuquerque, NM, and Lockheed Martin Corporation.)

The practical impacts of nanosensors will include having many sensors in a single system to improve both the range of materials that can be analyzed and the lifetime of the overall system.

In short, nanotechnology presents a historic opportunity for the improvement of existing sensors, as well as the development of new sensors that will have numerous new applications in many industries.

This chapter is based on an article, "Nanotechnology-Enabled Sensors: Possibilities, Realities, and Applications," that ran in the November 2003 issue of *Sensors* magazine (www.sensorsmag.com).

REFERENCES

1. Y. Cui, Q. Wei, H. Park, and C. M. Lieber, "Nanowire Nanosensors for Highly Sensitive and Selective Detection of Biological and Chemical Species," *Science* 293 (August 17, 2001): 1289–1292; "Space Mission for Nanosensors," *The FUTURIST,* November/December 2002, 13; J. A. Cas-

sell, "DoD grants $3M to study nanoshells for early detection, treatment of breast cancer," *NanoBiotech News* 1, no. 3 (August 13, 2003).

2. "Small Wonders, Endless Frontiers: A Review of the National Nanotechnology Initiative," *National Academy Press* (2002).
3. S. K. Moore, "U.S. Nanotech Funding Heads for $1 Billion Horizon," *IEEE Spectrum On-line,* http://www.spectrum.ieee.org/WEBONLY/resource/jun02/nnano.html.
4. "Nanosensors," Business Communications Company, Inc., 2004, http://www.marketresearch.com.
5. NanoMarkets LLC, "Nanosensors: A Market Opportunity Analysis," 2004, http://www.nanomarkets.net/news/pr_detail.cfm?PRID=156.
6. L. Bock, "Nano's Veteran Entrepreneur," *SMALLTIMES,* July/August 2003, 27–32.
7. http://www.research.ibm.com/resources/news/20020611_millipede.shtml.
8. I. Bernt et al., *Molecular Self-assembly: Organic Versus Inorganic Approaches,* 1st edition, ed. M. Fujita (Berlin: Springer Verlag, 2000).
9. www.zyvex.com/Capabilities/other.html.
10. Daan Frenkel and B. Smit, *Understanding Molecular Simulation,* 2nd edition (New York: Academic Press, 2001); and D. C. Rapaport, *The Art of Molecular Dynamics Simulation,* 2nd edition (Cambridge: Cambridge University Press, 2004).
11. R. Chandrasekaran, J. Miller, and M. Gertner, "Detecting Molecules: The Commercialization of Nanosensors," *Nanotechnology Law and Business Journal* 2, issue 1 (2005): 1–16. http://pubs.nanolabweb.com/nlb/vol2/iss1/1.
12. T. Vo-Dinh, B. M. Cullum, and D. L. Stokes, "Nanosensors and biochips: Frontiers in biomolecular diagnosis," *Sensors and Actuators B* 74 (2001): 2–11.
13. P. Poncharal, Z. L. Wang, D. Ugarte, and W. A. deHeer, "Electrostatic Deflections and Electromechanical Resonances of Carbon Nanotubes," *Science* 283 (1999): 1513–1516; and J. Toon, "Weighing the Very Small: 'Nanobalance' Based on Carbon Nanotubes Shows New Application for Nanomechanics," *Georgia Tech Research News,* Mar. 4, 1999.
14. A. N. Cleland and M. L. Roukes, "A Nanometre-scale Mechanical Electrometer," *Nature* 392 (March 12, 1998).

15. A. Modi, N. Koratkar, E. Lass, B. Wei, and P. M. Ajayan, "Miniaturized Gas Ionization Sensors using Carbon Nanotubes," *Nature* 424 (July 10, 2003): 171–174.

16. C. A. Grimes et al., "A Sentinel Sensor Network for Hydrogen Sensing," *Sensors* 3 (2003): 69–82.

17. J. Kong et al., "Nanotube Molecular Wires as Chemical Sensors," *Science* 287 (Jan. 28, 2000): 622–625.

18. K. Parikh et al., "Flexible Vapor Sensors Using Single Walled Carbon Nanotubes," *Sensors and Actuators B*, in press.

19. P. G. Datskos and T. Thundat, "Nanocantilever Signal Transduction by Electron Transfer," *J. Nanosci. Nanotech* 2, no. 3-4 (2002): 369–372.

20. W. L. Hughes and Z. L. Wang, "Nanobelts as nanocantilevers," *Applied Physics Letters* 82, no. 17 (April 28, 2003): 2886–2888.

21. R. Nicewarner-Pena et al., "Submicrometer Metallic Barcodes," *Science* 294 (2001): 137; and F. Freemantle, "Nano Bar Coding for Bioanalysis," *Science* 79, no. 41 (Oct. 8, 2001): 13.

22. E. Smalley, "Chip Senses Trace DNA," *Technology Res. News* (Jul 30/Aug. 6, 2003).

23. "Ultralight device analyzes gases immediately. Flying SnifferSTAR may aid civilians and US military," Sandia National Laboratories, press release, January 23, 2003.

Chapter 15

Microelectronics

The microelectronics industry has steadily improved the speed and efficiency of computing while simultaneously lowering costs in a highly predictable road map. This steady progression is popularly known as Moore's Law, after Gordon Moore, co-founder of Intel. Moore observed in 1965 that the number of transistors on a chip roughly doubles every two years.

Moore's Law has been the metronome of the microelectronics industry: The predictability of "faster-better-cheaper" allows the industry to precisely plan for the future in areas from engineering to capital expenditures to marketing. It has benefited everyone who is in the industry or purchases its products.

However, the physical limits of the current paradigm in computing are approaching, and the S-curve transition to the next stage is looming. The industry is driven by large companies that have a clear corporate focus and high equipment costs. They will not experiment with innovations from nanotechnology unless a clear, compelling case for their profitability can be made.

This chapter outlines manufacturing strategy, gives an overview of current technologies, and examines future devices we are likely to see. It concludes by examining photonics as a candidate for the technology that will enable continued progress at the lower limits of Moore's Law.

NANOMANUFACTURING PRODUCT STRATEGY

George Thompson

The use of nanotechnology in microelectronics is starting to take hold in the microelectronics industry, and this chapter discusses some of the strategic reasoning for incorporating nanotechnology into products that will be manufactured in high volume. This discussion begins with a review of the reasons any new technology should be incorporated into a new product and then discusses the specific case of nanotechnology.

A strategy for inserting a new technology into a product requires a clear understanding of both the existing product and the new technology if the manufacturer is to avoid the common case of a successful new technology becoming a commercial product failure. A clear understanding of the reasons for a product's success in the marketplace is critical before major changes are made in the product's base technology.

When an organization is evaluating innovations from nanotechnology and their success in the marketplace, there are a few points to consider. A key point is that most customers want to buy a superior product, and the technology embedded in that product is of secondary concern. Technology by itself is of little value to the customer unless it is incorporated into a product that people will buy.

A key to developing a coherent technology strategy for a new product is to consider whether the new technology creates a simple, one-time improvement opportunity or whether a rapidly evolving dynamic technology, such as nanotechnology, is involved. It is critical to understand the difference in the opportunities and the challenges. The incorporation of a static technology will be expected to provide a one-time change in the product, whereas the incorporation of a dynamic technology will create the possibility of improving the product in the near future as the base technology evolves.

Nanotechnology is a rapidly evolving technology. Predicting trends is difficult, and sudden major discontinuities in the technology are expected. This greatly complicates the business strategy for integrating nanotechnology into new products. It is helpful to consider a few simple but important points about the introduction of a static new technology into a product, without the added complications that exist in the case of a dynamic technology such as nanotechnology. A new technology can be embedded at several different stages in a product's life cycle, ranging from assumptions made in the early R&D stages up to the time when end users decide whether or not to buy the final product. Usually a series of highly objective processes is at work, although the role of irrational processes, such as fads, should not be ignored.

Consider the case of introducing a new technology, such as a new material, into an existing product. This new product will have to compete with the other products that serve the same or a similar function in the marketplace. The two most obvious factors to consider are performance and cost. Usually the customer base will quickly notice an improvement in price or performance and will transition to the new product if the improvement does not involve major design changes or integration investments. A product with superior performance, at the same cost, will have a very good chance of gaining market share over time, and in some cases will drive an increase in the total available market.

The case of a product being characterized by only price and performance is far too simple for practical use, although most products can be described by a reasonably small set of product characteristics—for example, function, features, performance, cost, reliability, delivery schedule, form factor (size), quantity (needed to succeed in the market), and input requirements (electricity, cooling, and so on). Other considerations include disposal costs and potential legal or ethical issues arising from any health or safety considerations related to the use of the product.

Before planning a significant change in a technology road map, the strategist who is planning to incorporate nanotechnology into a product line should begin by understanding the product's characteristics. The utility of a well-defined list of product characteristics is that it provides clear design guidance to the development team. In many cases the product characteristics are only vaguely defined, at the level of, "Is this supposed to be a sports car or a truck?" At times, a product is not defined even at this level, and the design team's primary goal becomes simply how to find an application for a new technology. Without a clear definition and without mapping customer requirements to the product's characteristics, the chance of an economically successful product launch decrease dramatically.

A review of the product's characteristics and of customer requirements—for example, the characteristics and customer requirements for a sports car versus those for a truck—makes for a useful exercise in clarifying ideas that are both simple and well known but often overlooked by designers when they incorporate new technologies. It is critical to focus on a product as a collection of product characteristics; a successful product is a specific collection of product characteristics that customers value enough to buy at a price that provides the manufacturer with a reasonable return on investment.

Considering Future Impacts

These comments on product characteristics serve to illustrate a few key ideas involved in launching a product based on nanotechnology. It's important to note that nanotechnology is dynamic and has an uncertain future. A critical long-term strategic concern is how nanotechnology will continue to evolve in the future and its overall impact on the microelectronics industry. In addition, planners must comprehend the future impacts on their products as nanotechnology continues to evolve.

A reasonable prediction of this process requires an understanding of how the specific nanotechnology affects the final product. It is also necessary to determine in some detail which product parameters will benefit, and which may be degraded, from the application of nanotechnology. Of course, this must be considered in enough detail to explain the impact of nanotechnology on competitors' products as well.

Ideally, the future evolution of nanotechnology should lead to a continuous improvement in the final product, an improvement that the customer will value. The best example of this type of long-term improvement is in the semiconductor industry, which has achieved unprecedented growth by providing its customers with products that are "faster-better-cheaper." This industry mantra is the key to creating new markets for its products and to making existing customers happy to replace old computers with newer ones as a result of a dramatic increase in performance.

The semiconductor industry has been able to maintain this scaling of better performance and better cost for a simple reason: Smaller transistors are both faster and cheaper. There is a natural synergy between these two key product parameters that is sufficiently strong to drive sales and increase the revenues needed to support additional investments in the technology R&D and infrastructure. This model permits the continuing investments needed to develop new and superior products. As mentioned earlier, this cycle, which results in the number of transistors doubling on a typical chip approximately every two years, is called Moore's Law. Its origin lies in the natural synergy that exists for semiconductor products between cost and performance as the size of the transistors is reduced, driving both the technology and the economics forward.

Identifying Potential Synergies

For the technologist who is considering the strategic impact of nanotechnology on a product, the first step may be to determine the potential synergies that exist between product parameters as a result of the introduction of nanotechnology. Which product parameters may be degraded by nanotechnology must also be understood. In addition to creating new applications and markets, the key is to identify potential synergies that will drive product improvement so rapidly that existing customers will want to upgrade to the newer technology.

One must always consider the risk that product acceptance may be driven by highly subjective factors. In the longer term, sustained market growth is more likely for products that successfully compete in the marketplace based on an unambiguously objective product superiority. The product should maintain compatibility with existing architectures, infrastructure, and product applications so that customers can easily incorporate the new products into their existing applications, in addition to possibly creating new applications.

Implicit in the definition of product characteristics is the ability to manufacture the new products in sufficient quantity on a schedule that meets the needs of customers, rather than the needs of a material or technology supplier. The manufacturing issues surrounding products that incorporate nanotechnology are complex because of the novelty of the technology. It is critical to ensure that a stable supply line to the customer is established, because a rational customer will frequently choose a reliable product that just meets the technology, cost, and schedule requirements over a potentially superior product that is not certain to be delivered on time.

Nanotechnology today presents a number of somewhat intimidating challenges to any technologist considering its use in a product. Given that nanotechnology is in its infancy, one must also consider that after nanotechnology is initially incorporated, the product may soon become obsolete as nanotechnology improves. This presents an opportunity to continuously improve the product in lockstep with the advances in nanotechnology, which in turn creates options for new markets. It also presents opportunities for competitors to leapfrog ahead of your product.

A clear understanding of the impact of nanotechnology on the final product characteristics, an awareness of the rapidly evolving nature of nanotechnology, and consistency with existing architectures, infrastructure, and metrics are all prerequisites for the successful formulation of a nanotechnology manufacturing strategy.

EXISTING TECHNOLOGIES

Stephen Goodnick

The exponential increase in the density of integrated circuits (ICs) predicted by Moore's Law has been driven primarily by shrinking the dimensions of the individual semiconductor devices comprising these circuits (Figure 15–1). Smaller device dimensions reduce the size of the circuits and therefore lead to a reduction of overall die area (the actual area partitioned on a silicon wafer corresponding to an individual integrated circuit), thus allowing for more transistors on a single die without negatively impacting the cost of manufacturing. However, getting more functions into each circuit generally leads to larger die size, and that requires larger wafers.

The workhorse of the semiconductor industry over the past three decades has been the metal oxide semiconductor field effect transistor (MOSFET). Its basic structure is comprised of two conducting regions (a source and a drain) separated by an insulating (oxide) gate over a channel that can be

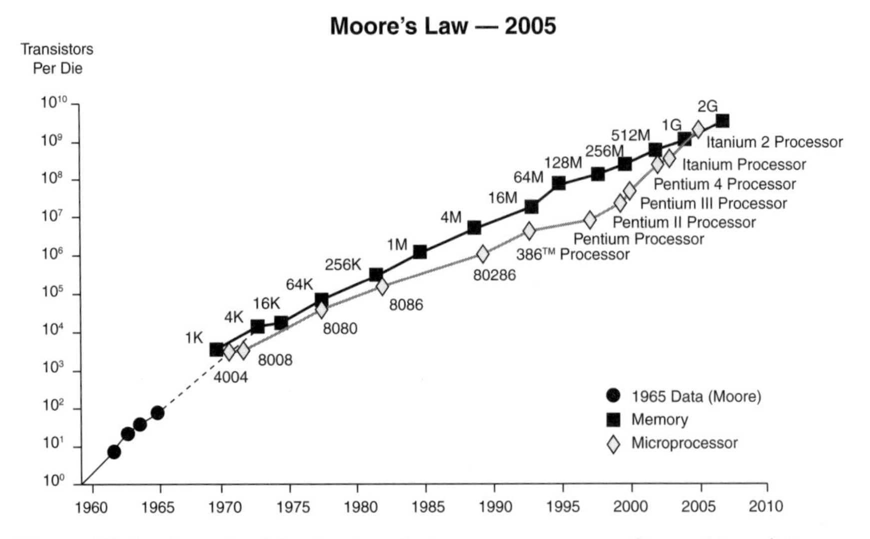

Figure 15–1 Growth of the density of microprocessors according to Moore's Law. The vertical axis is numbers of transistors per chip. (Reprinted by permission of the Intel Corporation; copyright Intel Corporation.)

turned on or off; this gate is a simple switch and constitutes the basic building block of computer architectures based on two-state logic.

The critical scale is the gate length, which corresponds to the distance between the source and the drain. As this length is reduced, all the corresponding dimensions of the device decrease in size or scale, according to well-defined scaling rules. In addition to reducing the area per transistor on the die, scaling down the gate length reduces the time it takes to move a charge (electrons) from the source to the drain, increasing the switching speed of the transistor and hence the clock speed at which logic circuits operate.

Successive generations of MOS transistor technologies are characterized in terms of this critical dimension, which, for present state-of-the-art commercial production, has already entered the nanometer (nm) scale dimension regime. Figure 15–2 shows scanning electron microscope photographs of Intel's current production transistors at the so-called 65-nm node, and successively shorter-gate-length devices realized in the research laboratory, down to 15nm. Clearly, present-day transistor technology is nanotechnology.

As semiconductor feature sizes shrink into the nanometer-scale regime, device behavior becomes increasingly problematic as new physical phenomena at short dimensions occur and as limitations in material properties are reached. Clearly, as we scale below 15-nm gate length, we eventually approach dimensions for which the channel is only a few silicon (Si) atoms long, at

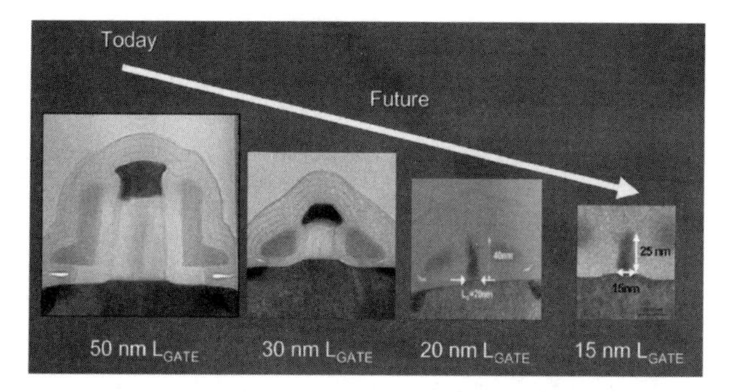

Figure 15–2 Scaling of successive generations of MOSFETs into the nanoscale regime. (Reprinted by permission of the Intel Corporation; copyright Intel Corporation.)

which point (or even much sooner) conventional MOSFET devices cannot scale further, and a saturation of Moore's Law will occur.

In fact, such a limit is already approaching in terms of the necessary silicon dioxide gate oxide thickness separating the gate metal from the channel, which, for 25-nm gate length technology, must be thinner than 1nm, as predicted by the International Technology Roadmap of Semiconductors (ITRS).[1] Basically, this reduction in thickness is required in order to maintain acceptable drive current, and to maintain charge control of the channel, by locating the gate as close to the channel as possible. This is because of the lateral length scale. One nanometer is only a few atomic layers thick, and for such a thin dielectric, leakage currents through the gate degrade performance and increase power dissipation. Industry is addressing this challenge by developing new dielectric materials with high permittivity, so that the effective gate capacitance is increased for a much thicker dielectric, giving better leakage performance.

Another issue leading to a saturation of Moore's Law is manufacturability as dimensions become smaller. As a semiconductor device becomes smaller, its output characteristics are increasingly sensitive to manufacturing and material imperfections. For example, impurity atoms with a valence higher or lower than Si must be introduced into the Si lattice at dilute concentrations to create free charges for carrying current; otherwise, the Si would be insulating. This process is referred to as "doping" the semiconductor, and in ultrasmall devices, the random position of dopant atoms in the device may cause dramatic changes in the current-voltage characteristics from device to device in very small structures. In a large device with many dopant atoms, this effect averages out, but in very small structures, the potential landscape seen by electrons traversing the source to drain varies widely from device to device because of the particular location of the dopant atoms.

Another source of device-to-device output characteristic variance is random variations in linewidths associated with lithography. Again, in very small devices, such process fluctuations may dominate the characteristics of a particular device. This sensitivity of nanoscale devices to process fluctuations means that not only manufacturing but also circuit and architecture design techniques must be developed to anticipate large device-to-device variations and even failures (fault-tolerant design).

To meet the challenges of shrinking gate lengths further into the nanometer-scale regime—increasing device performance and still maintaining control of charge in the channel with the gate—industry is moving away from the "classic" planar, bulk Si MOSFET design used during the past two decades. To increase performance for a given gate length, there is an increased trend toward alternative materials grown in the active regions of the device, such as strained Si, alloys of Si and Ge, and even compound semiconductors. Such alternatives to bulk Si increase performance through superior transport properties and hence faster switching.

To maintain charge as gate lengths shrink, research and production device structures are becoming increasingly nonclassical and three-dimensional rather than planar. These designs use Si on insulator technology (where a buried layer of oxide is introduced to isolate the device from the Si substrate), dual-gate and wraparound gates (in which the gate is above, below, and around the sides of the channel rather than simply on top), and nanowire shaped channels.

Figure 15–3 illustrates one such technology, the so-called FinFET (due to the fin-shaped gate). The structure is 3-D, and the channel resembles a 1-D wire with nanometer-scale dimensions (nanowire). This wirelike nature of the channel becomes important in the consideration of future technologies based on self-organized 1-D conductors such as carbon nanotubes (CNTs) and self-assembled semiconductor nanowires.

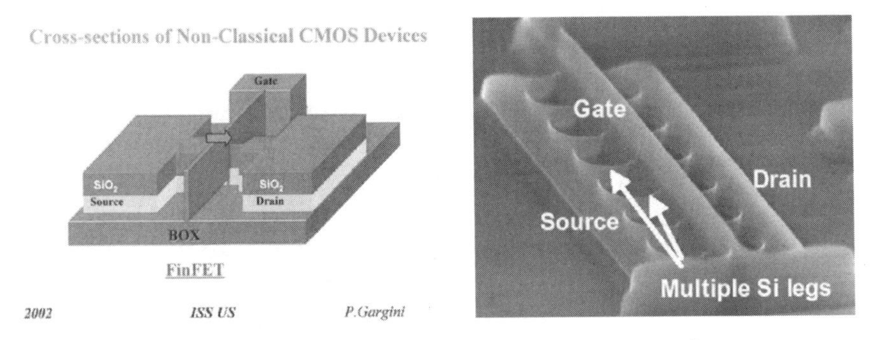

Figure 15–3 Nonclassical device structures. At left is a schematic of a FinFET; at right is an SEM photo of a multileg FinFET structure. (Reprinted by permission of the Intel Corporation; copyright Intel Corporation.)

Beyond these material and manufacturing issues, there are fundamental limits as device dimensions shrink. One is that quantum mechanics starts to play a role at small dimensions, in terms of the wavelike properties of charge carriers such as electrons. Effects such as quantization of motion, interference effects, and tunneling are all physical effects that modify the performance of devices at small dimensions.

Another limit is the discrete nature of charge. In small structures, charge can no longer be treated as a continuous fluid; rather, the number of charges is finite and small, leading to so-called single-electron charging effects. For very small structures, the change in energy and potential (voltage) due to one charge tunneling or otherwise moving from one conductor to another gives rise to a noticeable fluctuation in voltage. This sensitivity is because the capacitance (that is, the proportionality between charge and voltage, $\Delta Q = C\Delta V$) is a geometrical quantity that reduces as the structure size shrinks. If C is sufficiently small (10^{-17}F and less), then the change in voltage, ΔV, for a single electron moving from one side to the other ($\Delta Q = 1.6 \times 10^{-19}$C) may be larger than the thermal voltage, 25mV at room temperature.

All these effects can lead to noticeable degradation of the performance of classical and nonclassical MOSFETs, eventually leading to the end of the road map for scaling. Beyond that, there has been extensive work over the past decade related to nanoelectronic or quantum-scale devices, which actually use quantum mechanical and single-electron effects and operate on very different principles from conventional MOSFET devices. These alternatives may allow continued scaling beyond the end of the current scaling road map, as discussed later.

Future Nanoelectronic Device Technologies

As discussed earlier, as semiconductor device dimensions shrink to the nanoscale and beyond, the physics governing device behavior becomes complicated because of several factors. For large-dimension devices, the picture of macroscopic current flow in a device is analogous to fluid flow, in which charges and charge flow (current) appear continuous, and their motion is described by the same classical hydrodynamic equations used in the field of fluid mechanics. As we shrink to small dimensions, itís easy to see that at some level, charge is no longer continuous and that the motion of individual electrons becomes important in the behavior of nanoscale devices.

On the one hand, this sensitivity to individual unwanted charges threatens the reliability and reproducibility of nanoscale CMOS devices. On the other hand, the ability to control the state of individual electrons, and correspondingly to represent and store information, represents the ultimate limit of nanoscale device technology, which is the basis of so-called single-electron transistors and memory discussed later.

Another way in which the electronic behavior of small structures differs from that of macroscale systems is that electrons are governed by the laws of quantum mechanics, where matter exhibits both wavelike and particlelike behavior. One important length scale is the so-called De Broglie wavelength of an electron, which is the characteristic wavelength of matter waves in the quantum mechanics picture. The interaction of electrons with structures on this length scale resembles optics rather than classical dynamics, with effects such as diffraction, interference, quantization of motion, and tunneling, all of which lead to marked changes from the classical fluid picture of charge transport. Such wavelike behavior can persist over long dimensions, depending on the so-called phase coherence length, that is, the length over which an electron "wave" remains coherent.

Quantum computing is a new paradigm that explicitly depends on maintaining phase coherence, and using the potential information stored in the phase of a quantum mechanical two-state system, to exponentially extend the processing power compared with a simple binary logic system based on the same two states. Coherence is destroyed by the interaction of the electron with its energy-dissipative environment, primarily the vibrational motion of the host material in inorganic and organic structures. Because this vibrational motion increases with increasing temperature and thereby reduces the coherence length, quantum mechanical effects tend to wash out at room temperature. At room temperature, phase coherence lengths in Si, for example, are only a few tens of nanometers.

Generally, with regard to the behavior of conventional Si MOSFETs, single-charge and quantum mechanical effects adversely affect performance, creating barriers to further scaling at some future limit that is rapidly being approached, as discussed earlier. Beyond field effect transistors, however, there have been numerous proposals and demonstrations of device functionality and circuits based on single-electron and quantum mechanical effects. These include quantum interference, negative resistance, and single-electron devices, which are realized in metals, semiconductors, nanowires, carbon nanotubes, and molecular systems, as discussed in more detail later.

As dimensions become shorter than the phase-coherence length of electrons, the quantum mechanical wave nature of electrons becomes increasingly apparent, leading to phenomena such as interference, tunneling, and quantization of energy and momentum as discussed earlier. In fact, as was elegantly pointed out by IBM physicist Rolf Landauer, for a one-dimensional conductor such as a nanowire, the system is very analogous to an electromagnetic waveguide with "modes," each supporting a conductance less than or equal to a fundamental constant $2e^2/h$.

Such quantization of conductance was first demonstrated at Cambridge University and Delft University in the late 1980s, in specially fabricated, split-gate field effect transistors at low temperatures, where the split gate formed a one-dimensional channel in a field effect device. However, manifestations of quantized conductance, such as universal conductance fluctuations, noise, and the quantum Hall effect, appear in many transport phenomena. Many schemes were proposed for quantum interference devices based on analogies to passive microwave structures, such as directional couplers and even coupled waveguides for quantum computing. Promising results have been obtained on ballistic Y-branch structures by the research group in Lund, Sweden, where nonlinear switching behavior and elementary logic functions have been demonstrated, even at room temperature.

Most attempts at realizing quantum coherent devices suffer from the same problems as scaling of conventional nanoscale MOSFETs: the difficulty in controlling the desired waveguide behavior in the presence of unintentional disorder. This disorder can arise from the discrete impurity effects discussed earlier, as well as the difficulty of process control at true nanometer-scale dimensions. A further fundamental limit to devices based on the quantum mechanical nature of matter at the nanoscale is the phase coherence length and phase coherence time for maintaining a quantum coherent state. As mentioned earlier, this time and length scale is typically quite short in Si at room temperature.

In recent years, scientists have been attempting to exploit another purely quantum mechanical aspect of charge particles for nanodevice applications: that of electron spin. "Spin" refers to the intrinsic magnetic moment associated with elementary particles such as electrons, an effect that can manifest itself only through measurement relative to some particular reference frame in one of two states: spin-up or spin-down. This twentieth-century discovery has no classical analog, although the name itself implies an origin of magnetic moment due to a charge particle spinning around its own axis to generate a magnetic field.

In terms of the practical manifestation of spin, the ferromagnetic behavior of materials used in, for example, magnetic memory (a multibillion-dollar industry) is intrinsically associated with the interaction of spin states to form an ordered magnetic system. For nanoscale devices, the fact that there are two distinct states associated with spin has attracted researchers to the ability to encode information, either as simply binary information or as a prototypical "qubit" for quantum information storage. One of the main advantages of controlling the quantum state of spin is that spin is much more robust at preserving phase coherence compared with the ordinary quantum mechanical phase of an electron discussed earlier in connection with quantum interference-type devices. Typical spin coherence times in semiconductors can vary from nanoseconds to milliseconds, something that provides much more opportunity to realize quantum coherent devices for applications such as quantum computing.

Previously we mentioned the role of individual random charges as an undesirable element in the reproducibility of nanoscale FETs due to device-to-device variations. However, the discrete rather than continuous nature of charge of individual electrons at the nanolevel, and control of the motion of such electrons, is the basis of a great deal of research in single-electron devices and circuits. The understanding of single-electron behavior is most easily provided in terms of the capacitance, C, of a small tunnel junction (that is, two small conductors separated by a very thin insulator).

As mentioned earlier, capacitance is the proportionality constant relating the voltage difference between a pair of conductors to the net charge (positive on one, negative on the other) on the conductors; the simplest example is a parallel plate capacitor formed by two plates separated by an insulator. If a single electron tunnels across the thin junction from one side to the other, the change in net charge on the conductors results in a corresponding change in electrostatic energy, e^2/C. When physical dimensions are sufficiently small, the capacitance (which is primarily a geometrical) is correspondingly small, so that the change in energy may be greater than the thermal energy, $3/2kT$, resulting in the possibility of a "Coulomb blockade," or suppression of conductance due to the necessity to overcome this electrostatic voltage barrier. This Coulomb blockade effect allows the experimental control of electrons to tunnel one by one across a junction in response to a separate control gate, which can be used to lower this voltage barrier.

Figure 15–4 illustrates the operation of a so-called single-electron transistor, consisting of two tunnel junctions connecting to a conducting "island" or "quantum dot," to which a second voltage source V^g, is connected through a separate gate capacitor, C^g. As the gate voltage is increased, the Coulomb blockade is lifted when integer numbers of electrons tunnel through the structure, hence allowing control of electron motion one by one. Single-electron transistors, turnstiles, pumps, elementary logic circuits, and single-electron memories have been demonstrated experimentally, functioning even up to room temperature. Room-temperature operation is important for practical applications in that it does not require special cooling or cryogenic tech-

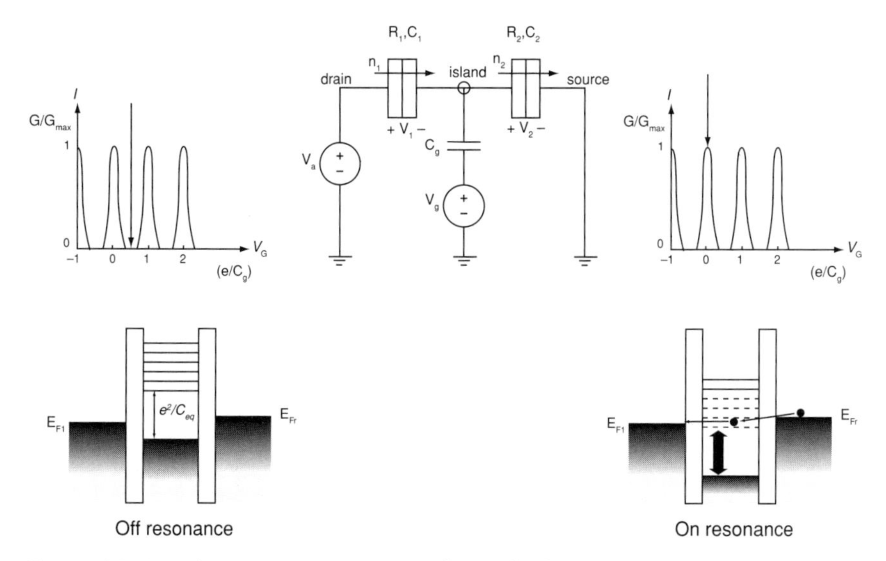

Figure 15–4 Schematic representation of a single-electron transistor (SET) consisting of two tunnel junctions connecting a conducting island, biased by drain-source voltage and controlled by gate voltage, V_g. During off resonance, electrons see a "gap" in energy, which prevents tunneling. Applying a voltage to the island through the gate allows electrons to tunnel one at a time, giving rise to a peak in conductivity.

nology, which would limit the applicability for, for example, portable electronics. As in the case of quantum interference devices, the technological difficulties arise from fluctuations due to random charges and other sources of manufacturing variation, as well as the difficulty in realizing lithographically defined structures with sufficiently small dimensions to have single-electron charging energies larger than the thermal energy, 25meV@300K.

There has been rapid progress in realizing functional nanoscale electronic devices based on self-assembled structures such as semiconductor nanowires (NWs) and carbon nanotubes. Semiconductor nanowires have been studied during the past decade in terms of their transport properties, and for nanodevice applications such as resonant tunneling diodes, single-electron transistors, and field effect structures.

Recently, there has been a dramatic increase in interest in NWs because of the demonstration of directed self-assembly of NWs via epitaxial growth. Figure 15–5 shows a scanning electron micrograph of such structures grown using vapor-liquid-solid epitaxy, where the dimensions of the nanowires are less than 10nm. Such semiconductor NWs can be elemental (Si,Ge) or III-V semiconductors, where it has been demonstrated that such wires may be controllably doped during growth, and abrupt compositional changes forming

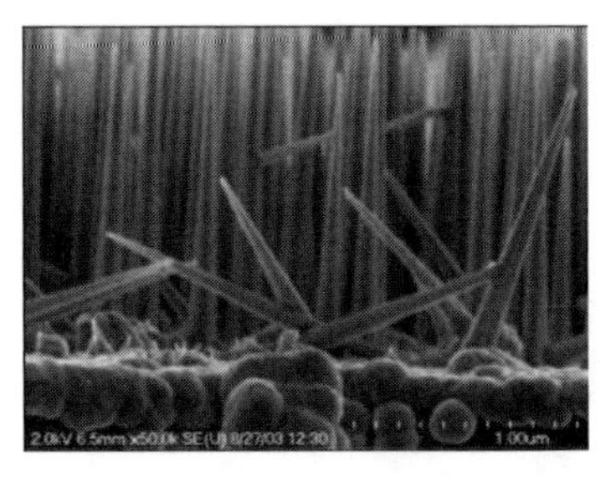

Figure 15–5 Scanning electron micrograph of self-assembled Si nanowires grown by vapor-liquid-solid epitaxy (after T. Picraux et al.).

high-quality -D heterojunctions can be achieved. Groups such as those at Harvard and Lund, Sweden, have demonstrated nanowire FETs, bipolar devices, and complementary inverters synthesized using self-assembly. The ability to controllably fabricate heterostructure nanowires has led to demonstration of nanoelectronic devices such as resonant tunneling diodes and single-electron transistors. The scalability of arrays of such nanowires to circuits and architectures has also begun to be addressed, although the primary difficulty is in the inability to grow and orient NWs with desired location and direction.

Carbon nanotubes are currently the focus of considerable attention because of the many remarkable properties of this new structural state of carbon. Figure 15–6 shows a schematic of a CNT that is composed of carbon atoms arranged in a stable tube configuration. It is a highly stable state of matter, very similar in concept to fullerenes like C60 (buckyballs). The structure can be envisioned as a graphite sheet (where the carbon atoms form hexagonal rings), which is rolled in a tube a few nanometers in diameter, as shown in Figure 15–6a. In rolling the tube and joining itself, the carbon rings forming the graphite structure can align in different offset configurations, characterized by their chirality. Depending on the chirality, CNTs can be metallic, semiconducting, or insulating, all the components required in conventional semiconductor IC technology (interconnects, transistors, and dielectrics). Field effect transistors have been fabricated from CNTs, and basic logic functions demonstrated by researchers at IBM and other research laboratories, as shown in Figure 15–6b. The extreme sensitivity of the conductivity of the nanotube to an attached atom or molecule to the wall or tip of the nanotube, also makes CNTs very attractive as sensors, the subject of considerable current research. The primary challenge faced in the evolution of this technology is the directed growth of CNTs with the desired chirality, and positioning on a semiconductor surface, suitable for large-scale manufacturing.

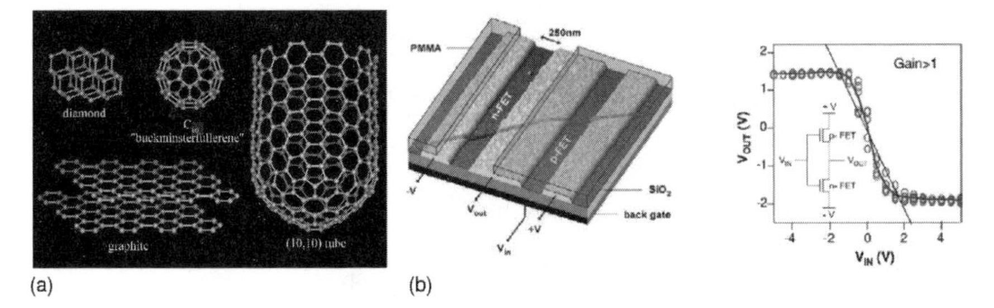

Figure 15–6 (a) Different states of carbon, including diamond, graphite, C60, and a carbon nanotube (right) (from Richard Smalley's image gallery, http://smalley.rice.edu/smalley.cfm); (b) carbon nanotube inverter formed from p- and n-channel FETs (from IBM, with permission).

Perhaps the ultimate limit of size scaling are devices comprised of a small number of molecules, forming the basis of electronic systems realized with molecular devices, or *molecular electronics* (moltronics). Figure 15–7 shows a schematic diagram of a nanoscale contact to a molecular device, through which current is passed. Here the molecular device is an organic chain to which different side groups or molecules are attached to realize a desired functionality. The molecular chain structure shown in the lower half of the figure, studied by Mark Reed (Yale) and James Tour (Rice), showed "negative differential conductance (NDC)" in the current voltage characteristics, that is, a decreasing current with increasing voltage. From a circuit standpoint, NDC appears as a negative resistance, which leads to signal amplification and the possibility of bistable behavior because the circuit does not like to reside in the regime, which is the basis for elementary switching devices. Elementary molecular electronic architectures have been demonstrated by HP Research Laboratories using crossbar-type logic.

A very attractive feature of molecular systems is the possibility of bottom-up or self-assembly of functional systems. Such templated self-assembly is of course the basis of biological systems, which have exquisite complexity

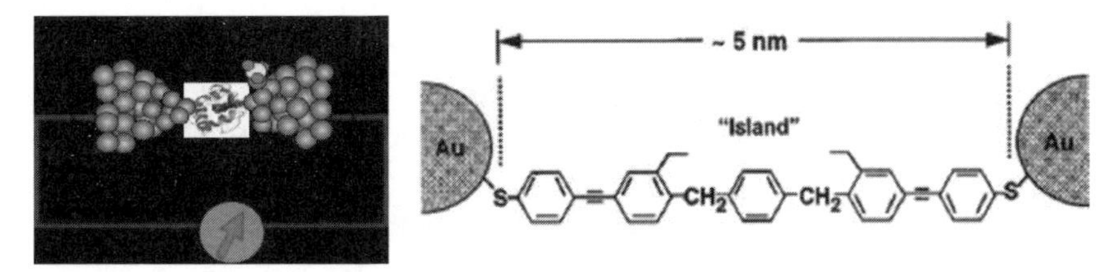

Figure 15–7 A molecular "junction" (left), and the corresponding molecular device contacted by external leads.

and functionality as well as self-replication and self-repair. Such "biomimetic" approaches to molecular circuits would represent an inexpensive alternative to the exponentially increasing cost of top-down nanofabrication, which is currently driving fab costs into the billions of dollars. However, at present there is no clearcut manufacturing approach to self-assembly in the near term.

Another difficulty in understanding and using molecular electronic structures is the need to separate the intrinsic behavior of a molecular device from the contacts themselves. In conventional devices, contacts are nearly ideal, providing a connection to the other devices and the external world through interconnects, and not affecting the intrinsic performance of devices except through well-controlled parasitic contact resistances. As the number of devices per chip scales exponentially, the number of contacts and interconnects per device increases even faster, and from an architecture standpoint, system performance is increasingly dominated by the interconnects and not the devices themselves. In a nanoscale devices, the contacts may in fact dominate the performance of the device, and at a minimum they are an integral part of the device. This problem is particularly evident in molecular electronic devices, as the schematic of Figure 15–7 indicates (where the contact is much larger than the device). This challenge remains one of the many issues to be solved in evolving molecular electronics in the future.

REFERENCE

1. International Technology Roadmap of Semiconductors, http://public.itrs.net/.

PHOTONICS

Axel Scherer

Photonics has recently become an attractive alternative to electronics for communications and information processing. Devices that use photons rather than electrons as information carriers can benefit from higher speeds and reduced cross talk between optical channels. Miniaturization of compact optical components such as resonators, waveguides, and interferometers has become very desirable. At the same time, microfabrication has emerged as a powerful technology that enables the construction of sub-100nm structures in a reproducible and controllable manner. The same technology that was driven by the continuing desire to miniaturize electronic components on silicon microchips has now evolved to a precision that allows us to control the flow of photons.

Fully optical chips would deliver the ultimate in speed and would benefit from lower heat and fewer power problems. A significant bottleneck in current networks is the switching from optical to electrical media for processing and routing, and then back from electrical to optical. If this bottleneck can be reworked to smoothly avoid the OEO transformation, tremendous gains can be attained. The need for data transfer and routing with high bandwidth is also compelling for many military applications, where optical solutions to high-frequency signal processing and routing offer lower power dissipation and higher bandwidth, which produce more robust and compact systems. Size, weight, and immunity to radio-frequency interference are particularly important for mobile communications, sensors, and radar systems.

Recent technological advances in silicon optoelectronics have highlighted the need for inexpensive multiwavelength light sources on silicon photonics. The use of silicon CMOS electronics for the multiplexing, control, and routing of many parallel optical signal channels on a silicon chip may have been an inevitable outcome of the lack of adequate electronic interconnection solutions for next-generation microprocessors. Silicon on insulator (SOI) waveguiding optics may produce much less expensive alternatives to more conventional telecommunications and data communications platforms such as GaAs, InP, and Lithium Niobate ($LiNbO_5$).

The most important disadvantage of using silicon optoelectronics has been the lack of sufficient gain for signal amplification and efficient on-chip light generation. The vision of an on-chip silicon laser has so far been elusive, although much time and effort have already been invested in various promising silicon light-amplification strategies, such as porous silicon and erbium-doped silicon waveguides.

Silicon nanophotonic components are so small that components for 1,000-wavelength division multiplexing channels could easily fit in a corner of an electronic chip. Optical nanodevices can now be constructed from the standard electronic semiconductor materials, such as silicon on insulator (SOI), Gallium Arsenide (GaAs), and Indium Phosphide (InP). By combining the need for integrated photonics with the capabilities offered by high-resolution microfabrication, the field of nanophotonics has emerged. Optical devices that have traditionally been constructed of glass and lithium niobate can now be scaled down in silicon, GaAs, or InP. Ultrasmall optical systems can also be integrated, thus realizing for the first time the dream of large-scale, multifunctional, all-optical chips for information processing. Moreover, because nanophotonics devices are constructed from standard electronic materials, such devices can be integrated side by side with electronic components, enabling the construction of hybrid systems of higher complexity.

The emergence of silicon nanophotonic technology in SOI wafers has made a profusion of optical components available on chip and essentially at zero marginal cost. These components include resonators, filters, waveguides, modulators, and (with the availability of germanium) detectors. To these must be added the full functionality of CMOS electronic technology, particularly high-quality transistors.

Thus there would be a great advantage in a mode-locked source that provides steady output at many frequencies simultaneously. In passive optical filtering and routing chips, waveguide losses, insertion losses, and resonator losses all contribute to a deterioration of the input signal. Thus, it is desirable to include gain in such chips so that the routed signal is amplified before coupling out of the routing switch. Good-quality mode locking requires high-Q cavities and a good optical modulator, both functions that are now readily fabricated on an SOI silicon chip. Thus many of the ingredients for highly dense microoptical circuits and for an internal mode-locked light source in silicon SOI are already available—except for optical gain. A goal, then, is the large-scale integration of multiwavelength sources within a single integrated chip.

Recently, optical coupling to disk and ring resonators has become an extremely effective way of fabricating add/drop filters. Much of the work has been demonstrated in glass waveguides, both monolithically and through micromechanical coupling of microspheres close to thinned-down optical fibers. An 8×8 crossbar router was recently demonstrated by Little et al. in a planar waveguide geometry in which high-index glass disks were aligned above the waveguide layers. Very high Qs and correspondingly narrow spectral filters have been demonstrated.

The minimum feature size required to couple the resonator disk to the waveguide, approximately 150–500nm, can already be obtained with high-resolution lithography, UV lithography, and electron beam lithography. Other lithography techniques, such as embossing, imprinting, and molding, will undoubtedly follow for the lower-cost development of high-resolution single-level lithography.

The photonic crystal (PC) is one of the platforms that can enable the miniaturization of photonic devices and their large-scale integration. These microfabricated periodic nanostructures can be designed to form frequency bands (photonic band gaps) within which the propagation of electromagnetic waves is forbidden, irrespective of the propagation direction. One of the most attractive planar photonic crystal devices is a compact and efficient nanocavity. This is due to the extraordinary feature of Planar Photonic Crystals

(PPCs) to localize high electromagnetic fields into very small volumes for a long period of time. Moreover, PC cavities can be engineered to concentrate light in the air, and thus they are natural candidates for the investigation of interaction between light and matter on a nanoscale level. Such nanoscale optical resonators are of interest for a number of applications, of both practical and scientific importance.

Ultrasmall quantities of biochemical reagents can be placed in the air region where field intensity is the strongest, and their (strong) influence on the optical signature of the resonator can be monitored. This can lead to realization of integrated spectroscopy systems (for example, on-chip Raman spectroscopy). PC nanocavities can have high Q factors (>10,000) and can be highly integrated (less than 5μm apart), something that makes them promising candidates for realization of channel drop filters in dense wavelength-division multiplex systems.

CONCLUSION

The microelectronics industry, and with it a portion of the U.S. economy, faces a critical hurdle as the physical limits of CMOS technology draw ever closer. Moore's Law has been critical to its rapid growth as both a metronome and a road map, but an S-curve transition to molecular electronics may create a plateau more than a decade from now in the steady improvement seen over the past decades. During the transition, technologies like photonics and new, fault-tolerant architectures will still increase computing efficiencies, although likely not at a pace commensurate with Moore's Law.

Chapter 16

Drug Delivery

Jianjun Cheng and Suzie Hwang Pun

Therapeutic drugs are usually delivered in bolus doses, either orally or by injection. These administrations result in initial blood concentrations that are higher than required levels, followed by a decrease to subtherapeutic levels due to drug degradation and excretion. Therefore, drugs must be given frequently to maintain therapeutic drug concentrations (Figure 16–1). In the 1970s, researchers introduced the concept of controlled drug delivery by using carriers to release drugs in a sustained manner. In the ideal case, drugs are gradually released from a depot so that the drug concentration is maintained

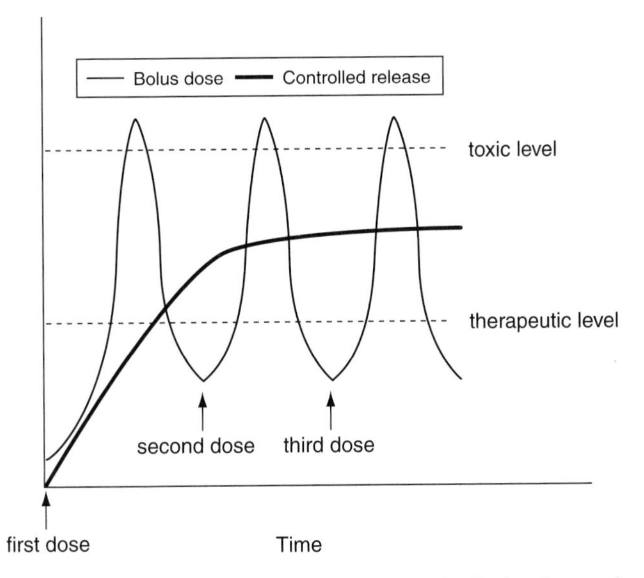

Figure 16–1 Schematic expression of plasma drug level of bolus dose and controlled release.

at an effective level over a long period. An example of one such success is Gliadel, the first FDA-approved biopolymer drug-delivery system for treatment of brain cancer. Gliadel wafers are implanted into the brain after tumor resection. There, the wafers locally release carmustine (a chemotherapeutic drug) for several months.

Many other promising drugs never make it to clinical trials because of inherent pharmacological drawbacks. Low-molecular-weight drugs, such as most chemotherapy drugs, are usually insoluble and highly toxic. Protein and nucleic acid drugs usually have poor stability in physiological conditions. It is therefore essential for these drugs to be protected en route to their target disease sites in the body. Drug-delivery systems may rescue potential drug candidates by increasing solubility and stability.

Drug-delivery technologies are developed to improve the safety and efficacy of drugs, to ensure better patient compliance, and to improve the shelf life and stability of therapeutic products. Controlled drug release involves the combination of a biocompatible material or device with a drug to be delivered in a way that the drug can be delivered to and released at diseased sites in a designed manner.

The major routes of drug administration are oral, inhalation, injection, and transdermal delivery. The most well-known route is oral drug delivery, which accounted for about 50 percent of the market as of 2003. The other routes of administration—inhalation, transdermal, injection and implantation, and nasal delivery—account for the remaining market share at 19 percent, 12 percent, 10 percent, and 7 percent, respectively. In the past 30 years, the field of drug delivery has been undergoing rapid development and has attracted attention from both academia and pharmaceutical industries. According to a recent report, the U.S. market alone for drug delivery is estimated at $43.7 billion in 2003 and is expected to grow more than 11 percent annually in the next five years.[1]

NANOTECHNOLOGY IN DRUG DELIVERY

Nanotechnology, a multidisciplinary area of science involved in the miniaturization and use of materials or devices on the nanometer scale, has been undergoing explosive growth and has become a principal research interest. In the past ten years, nanotechnology has already demonstrated great impact in almost every frontier area of drug delivery by extending and improving traditional delivery techniques.

One of the earliest applications of nanotechnology in drug delivery was conceived in the 1970s, when nanoparticles were designed as carriers of anticancer drugs. Since then, many nanoparticle systems have been developed for use in drug delivery. Biopolymers have been intensively studied for application in nanoparticulate drug delivery.

Delivery vehicles involved with polymeric systems include polymer-drug conjugates, polymeric micelles, polymeric nanospheres and nanocapsules, and polyplexes. Inorganic and metallic materials have also been used in preparation of nanoparticles for drug delivery. Recently nano- and microfabrication technologies have been applied for drug delivery, resulting in novel devices such as biochips and microneedles. With the explosive growth in the development of new methods of nanofabrication, numerous emerging nanosystems will inevitably change the field of drug delivery in the coming decades.

Nanoparticles for Drug Delivery

Nanoparticle-based delivery vehicles improve drug efficacy by modulating drug pharmocokinetics and biodistribution. Small-molecule drugs are rapidly eliminated from the circulation by the kidneys. Injectable nanoparticle-delivery vehicles, typically ranging from 5nm to 200nm in size, substantially increase circulation (particles >5nm avoid kidney clearance) while minimizing removal by cells that police the blood for foreign particles (macrophages have less propensity for particles <200nm in size). Oral delivery is currently the most preferred method of drug administration because of its cost effectiveness and ease of use.

The market for oral drug-delivery systems has been growing at a rate of 8.6 percent per year since 2000. A major area of research in oral delivery is in delivery materials for protein drugs. Because particle permeability across the intestinal wall is inversely proportional to size, nanoparticles used for oral delivery offer obvious advantages. The interest in nanoparticle-based drug delivery for other administration routes is also growing. The following sections focus on polymer, lipid, and inorganic or metallic nanoparticles that are <500nm in size.

Polymer Conjugates (Polymer-Drug Chemical Linkage)

Polymer-drug conjugates (5–20nm) represent the smallest nanoparticulate delivery vehicles. The polymers used for such purposes are usually highly water-soluble and include synthetic polymers (for example, poly(ethylene glycol) (PEG)) and natural polymers (such as dextran). When hydrophobic small molecules are attached to these polymers, their solubilities in water can be substantially improved. For example, a cyclodextrin-based polymer developed at Insert Therapeutics increases the solubility of camptothecin, an insoluble chemotherapy drug, by three orders of magnitude.

Small molecules or proteins conjugated to these polymer delivery vehicles can achieve extended retention in circulation because of reduced kidney clearance. PEG-L-asparaginase (ONCASPAR; Enzon), an FDA-approved PEGylated protein drug as a treatment for acute lymphoblastic leukemia, can be administered every two weeks, instead of the two to three times per week required for the non-PEGylated enzyme. Other PEGylated systems approved by the FDA include PEG–adenosine deaminase (ADAGEN; Enzon) as a treatment for X-linked severe combined immunogenicity syndrome and PEG-interferon (PEGASYS; Roche and PEG–INTRON; Schering-Plough) as treatments for hepatitis C.[2] Many polymer-small molecule and polymer-protein conjugates are currently in clinical trials. A promising nanoparticle drug-delivery system—a proprietary, albumin-bound paclitaxel conjugate codeveloped by American Pharmaceutical Partners and American BioScience—has shown excellent antitumor efficacy in Phase III clinical trial for the treatment of metastatic breast cancer.

One group of polymers that has attracted enthusiasm recently are dendrimers (Figure 16–2). They are monodispersed, symmetric, globular-shaped macromolecules comprising a series of branches around an inner core. Dendrimers are potential nanometer-sized systems for drug delivery, and their

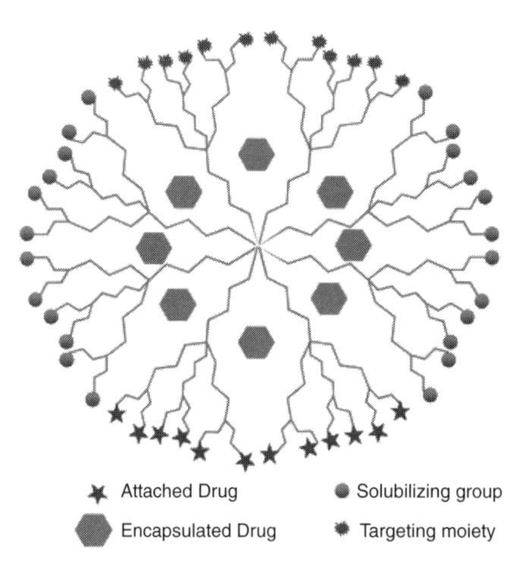

Figure 16–2 Schematic drawing of dendrimer for application in drug delivery and targeting. (Reprinted with permission from http://www.drugdeliverytech.com/cgi-bin/articles.cgi?idArticle=153, Thiagarajan Sakthivel, Ph.D., and Alexander T. Florence, Ph.D., D.Sc., "Dendrimers & Dendrons: Facets of Pharmaceutical Nanotechnology.")

sizes can be controlled simply by adjusting the generation of dendritic branches.

Polymer Micelles (Polymer Surfactant-Drug Self-Assembly)

Amphiphilic block copolymers—polymers that contain both hydrophilic and hydrophobic regions—tend to self-assemble in aqueous solution into spherical structures called micelles. Polymeric micelles typically have a hydrophilic corona and a hydrophobic shell. When used as drug-delivery agents, polymeric micelles are most commonly formulated to include hydrophobic drugs (for example, doxorubicin, cisplatin, amphotericin B) in the core, leaving the outer hydrophilic layer to form a stable dispersion in aqueous media.

The stable structure of the polymeric micelles prevents rapid dissociation (release of drug) in vivo. Polymer micelles typically range from 60 to 100nm, with fairly narrow size distributions. The micellar corona can be further modified with targeting moieties (for example, antibodies) to deliver drugs to desired sites. Development of micellar drug-delivery vehicles is in an early stage, with most formulations still in preclinical studies.[3]

Polymer Nanoparticles (Drug Dispersion or Encapsulation in Polymer Aggregates)

Nanoparticles are solid, small colloidal particles made of polymers having diameters from 50nm to several hundred nanometers. Depending on the method of preparation, two types of drug-containing nanoparticles exist: nanospheres (a matrix system in which drugs are uniformly distributed) or nanocapsules (a reservoir system in which drugs are confined to the core of the particles and are surrounded by polymer membranes). Many of these systems are made of biodegradable polymers, such as poly(ortho ester) (Figure 16–3). Poly(ortho ester) nanoparticles release drugs with tunable release rates, depending on solution pH. Drugs encapsulated in nanoparticles have increased stability against enzymatic and chemical degradation, an important advantage for unstable drugs such as proteins and nucleic acids.[4]

Nanoparticles have been tested for the delivery of all types of drugs (small molecules, proteins, and nucleic acids) in almost all types of administration routes (such as inhalation, oral, and injection). Although many of these approaches are still in an early stage of their development, some of them have already shown great potential. An example is Dr. Edith Mathiowitz's (Brown University) poly(fumaric-co-sebacic) anhydride nanoparticles for the oral delivery of insulin, a promising way to achieve oral protein delivery.[5]

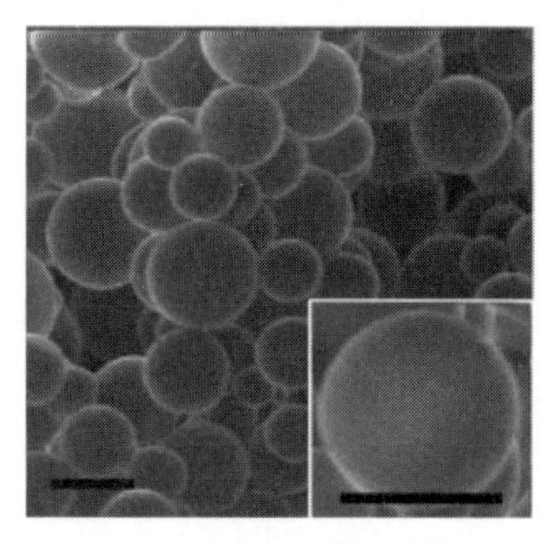

Figure 16–3 Scanning electron microscopy image of poly(ortho ester) nano- and microspheres. Scale bars: 5mm. (Reprinted with permission from Chun Wang, Qing Ge, David Ting, David Nguyen, Hui-Rong Shen, Jianzhu Chen, Herman N. Eisen, Jorge Heller, Robert Langer, and David Putnam, "Molecularly engineered poly(ortho ester) microspheres for enhanced delivery of DNA vaccines," *Nature Materials* 3 (2004): 190–196.)

Polyplexes (Polymer-Nucleic Acids Complex through Charge Interaction)

In nonviral gene therapy, plasmid DNA is introduced into cells to express therapeutic proteins, whereas in oligonucleotide therapy, oligonucleotides (such as ribozymes and DNAzymes) and small, interfering RNA (siRNA) are used to suppress disease-associated expression. However, the cell membrane is a natural barrier for these genetic materials. For therapeutic nucleic acids to be successfully delivered into the cell, they must be complexed with materials that facilitate cellular uptake.

Polyplexes, a group of nanoparticulates formed by charge interaction between positively charged polymers and negatively charged nucleic acids, are developed for such purposes. Polyplexes range in size from 40nm to 200nm (Figure 16–4). RNA interference (RNAi) is an emerging and promising approach for oligonucleotide therapy, and there is currently active research in developing materials for siRNA delivery.

Liposomes

Liposomes—nano-sized particles (25 to several hundred nanometers) made from phospholipids and cholesterols—are sometimes referred to as "fat bubbles." Liposomes consist of bilayers of lipids that can encapsulate drugs. Their properties for use as delivery vehicles are closely associated with lipid composition, liposomal size, and fabrication methods. For example, saturated phospholipids with long hydrophobic chains usually form a rigid, impermeable bilayer structure, whereas the unsaturated phosphatidylcholine-based lipid layers are much more permeable and less stable.

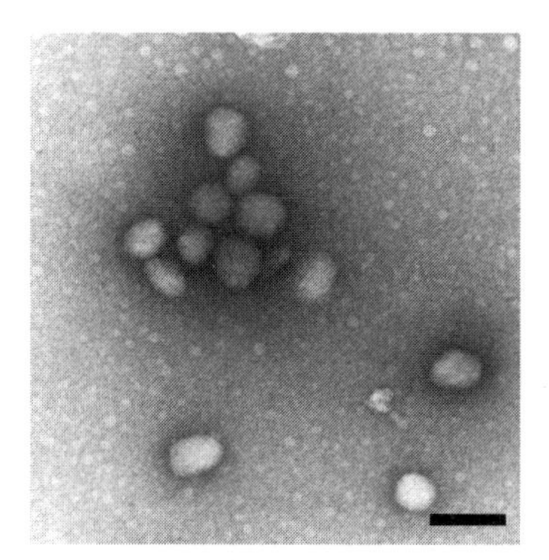

Figure 16–4 Cyclodextrin polycation-based polyplexes developed by Davis and coworkers (California Institute of Technology).[6] These materials are being investigated for gene therapy applications at Insert Therapeutics. Bar is 100nm. (Reprinted with permission from S. J. Hwang, N. C. Bellocq, and M. E. Davis, "Effects of Structure of beta-cyclodextrin-containing Polymers on Gene Delivery," *Bioconjugate Chemistry* 12(2) (2001): 280–290.)

Liposomal drug delivery has achieved great success in the past decade. Several liposome-based drug-delivery systems have been approved for clinical use. AmBisome (lipid-based delivery of amphotericin B, Fujisawa Healthcare, Inc., and Gilead Science) was approved for the treatment of cryptococcal meningitis in HIV-infected patients. The sales of AmBisome were nearly $200 million in 2003 (an increase of 7 percent from 2002). Doxil (Alza) was approved in 1999 for the treatment of refractory ovarian cancer and is the first and only liposomal cytotoxic agent approved to treat a solid tumor. The sales of Doxil reached $80 million in 2000.

Inorganic and Metallic Nanoparticles

Delivery of drugs using new inorganic and metallic nano-sized vectors are still in the proof-of-concept of stage. One unique approach originates from C-60, a soccer ball–shaped fullerene.[7] Another interesting approach is to use magnetic nanoparticles to carry chemotherapeutic drugs to cancer sites directed by an external magnetic field.

Very recently, other metal nanoparticles have been investigated as therapeutics and drug-delivery systems. An example from Dr. Naomi Halas's research group (Rice University) is the nanoshell, a new type of nanoparticle composed of a dielectric silica core coated with an ultrathin gold layer.[8] Once the nanoshells penetrate tumor tissues, they can be activated for thermal therapy by taking advantage of their ability to convert absorbed energy from the near-infrared region to heat.

Implantable Drug Delivery

Implantable drug delivery (IDD) devices offer more uniform drug release rates, lower required doses, and localized delivery. These devices are generally composed of the drug of interest distributed in a polymer matrix. Examples of marketed implantable drug-delivery formulations include Norplant for birth control (Wyeth Laboratories), Gliadel for localized delivery of a chemotherapeutic agent to the brain (Guilford Pharmaceuticals), and Viadur for slow release of hormones for prostate cancer treatment (Bayer). The sales of Gliadel in 2003 were nearly $20 million (an increase of 32 percent from 2002), and annual Viadur sales are projected to reach $150 million.

Based on these trends, the demand for implantable drug-delivery systems is expected to exceed $2 billion by 2012. There is still room for improvement in IDD technology. Two emerging nanotechnologies with applications in implantable drug delivery are nanoporous membranes and biochips.

Nanoporous Membranes

Nanoporous membranes are microfabricated with well-defined pores (diameters in the tens of nanometers). The membranes can be used to deliver small-molecule, peptide, or protein drugs (Figure 16–5). One application under investigation involves encapsulation of pancreatic islet cells for insulin delivery. The reproducible and uniform pore size precisely controls the material exchange across nanoporous membranes: Nutrients for the cells and secreted insulin can pass through the pores, but proteins and cells from the immune system that may attack the implanted islet cells are restricted from entering the biocapsules due to their size.

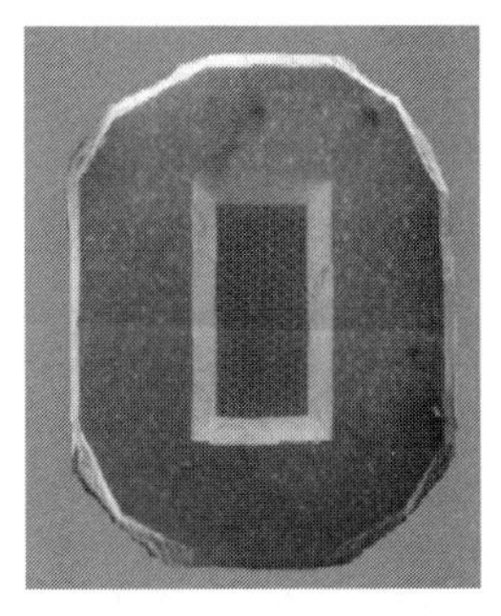

Figure 16–5 Nanoporous membranes developed by Desai and colleagues have nanometer-sized pores for controlled material exchange.[9] (Reprinted with permission from S. L. Tao and T. A. Desai, "Microfabricated drug delivery systems: From particles to pores," *Advanced Drug Delivery Reviews* 55 (2003): 315–328.)

Biochips

Drug release from IDD devices is not constant, and "burst" effects are still observed. In addition, drug release cannot be controlled after implantation of IDD devices. Biochips have been developed to precisely control the amount of drug released. Biochips are usually fabricated using silicon and contain a number of mini-wells with precisely controlled sizes, each capable of holding a few hundred nanoliters. These mini-wells are loaded with drugs and are covered with caps of thin metal foils (usually gold) that are connected to the wires on the face of the chips (Figure 16–6).[10] When electrical signal is applied, the current dissolves the metal covers and releases the drug. Biochips can be implanted beneath the skin or into more specific areas such as the spinal cord or brain. The electronics package outside the chips receives a radio-frequency instruction through a built-in antenna to order a microprocessor to control the melting of metal foils. MicroCHIPS, Inc., is one of the key companies developing this technology. Preclinical studies in animals using biochips developed by MicroCHIPS, Inc., have shown good biocompatibility without significant side effects. Once successfully developed, the biochip-based drug-delivery technology will allow for precisely controlled drug administration to patients.

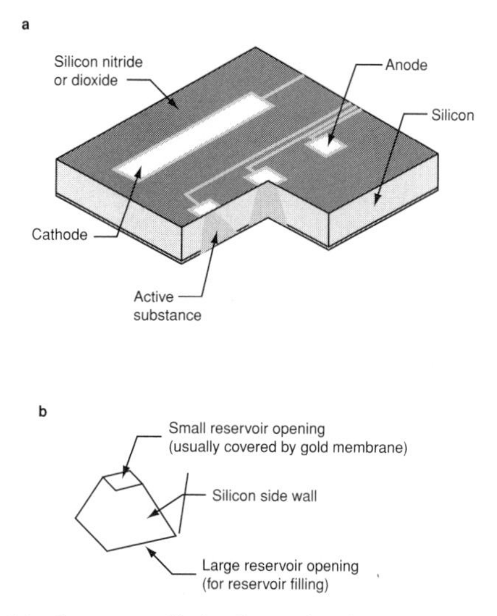

Figure 16–6 A biochip for controlled release developed by Robert Langer and coworkers. (a) Schematic drawing of biochip; (b) a single reservoir. (Reprinted with permission from J. T. Santini, M. J. Cima, and R. Langer, "A controlled-release microchip," *Nature* 397, no. 6717 (Jan. 28, 1999): 335–338.)

Transdermal Drug Delivery

The first transdermal patch technology for drug delivery was approved by the U.S. FDA in 1979. In the past quarter-century, the U.S. market for transdermal patches has grown to more than $3 billion per year.[11] Transdermal drug delivery is an attractive, noninvasive approach to drug administration and is likely to improve the bioavailability of drugs by avoiding first-pass liver metabolism and by maintaining more uniform drug plasma levels than bolus injections.

Despite these advantages, successful application of transdermal patches has been largely limited to a handful of drugs, including nicotine (for smoking cessation), scopolamine (for motion sickness), and hormone replacement. The human skin acts as a protective barrier to prevent entry of most molecules. A lipid-rich layer in the skin repels water-soluble molecules, and large molecules are also largely excluded because of their size. Thus, most major families of drugs (hydrophilic small molecules, peptides, proteins, and nucleic acids) have not been successfully delivered by traditional transdermal patch technology.

The permeability of the skin to hydrophilic or macromolecule drugs can be increased by physically creating microscopic holes in the skin using arrays of microneedles. Ideally, the microneedles would create pores to facilitate drug entry but would remain small enough to be painless. Although technology was conceived several decades ago, the technology to prepare these systems was not available. Microfabrication technology, developed by the microelectronics industry, recently has been used to prepare these structures.

Prausnitz and colleagues first demonstrated this concept in 1998 by preparing silicon-based, solid microneedles (Figure 16–7). Since then, many

Figure 16–7 Solid silicon microneedles with heights ~150nm can be applied painlessly to enhance transdermal drug delivery. (Reprinted with permission from M. R. Prausnitz, "Microneedles for transdermal drug delivery," *Advanced Drug Delivery Reviews* 56, no. 5 (2004): 581–587.)

variations of these structures have been prepared and tested with diverse drug families. Materials used to synthesize microneedles include glass, polymers, metal, and silicon. The microneedles have been successfully applied to deliver small molecules, oligonucleotides, plasmids, and proteins through the skin, increasing the permeability of these molecules by several orders of magnitude. Application approaches include using solid microneedles to pierce holes in the skin, followed by application of a traditional patch; coating solid microneedles with drugs that are slowly released from the needles after skin penetration; and flowing drug solutions through hollow microneedles.

A pilot human trial revealed that the microneedle arrays are indeed applied painlessly.[12] Transdermal microneedle delivery was quickly adopted by the pharmaceutical and drug-delivery industries; several companies are currently developing this technology for transdermal drug administration. As this technology matures, it has the potential to quickly surpass the market currently occupied by traditional patch formulations.

FUTURE TRENDS IN DRUG DELIVERY

Nanotechnology has played a large role in advancing the drug-delivery field by enhancing existing areas of small-molecule and protein delivery and by opening doors for delivery of new families of nucleic acid-based drugs. The ability to control the properties of nanoscale materials will continue to impact the pharmaceutical field by providing new technologies for improved drug delivery. We are optimistic that these developments will lead to delivery vehicles with high target specificity and with the ability to precisely control drug release.

REFERENCES

1. O. Gorka, M. R. Hernández, A. Rodríguez-Gascón, A. Domínguez-Gil, and J. L. Pedraz, "Drug delivery in biotechnology: Present and future," *Current Opinion in Biotechnology* 14 (2003): 659–664.
2. R. Duncan, "The dawning era of polymer therapeutics," *Nature Reviews Drug Discovery* 2 (2003): 347.
3. T. Nakanishi, S, Fukushima, K. Okamoto, M. Suzuki, Y. Matsumura, M. Yokoyama, T. Okano, Y. Sakurai, and K. Kataoka, "Development of the polymer micelle carrier system for doxorubicin," *Journal of Controlled Release* 74 (2001): 295–302.

4. C. Wang, Q. Ge, D. Ting, D. Nguyen, H. Shen, J. Chen, H. N. Eisen, J. Heller, R. Langer, and D. Putnam, "Molecularly engineered poly(ortho ester) microspheres for enhanced delivery of DNA vaccines," *Nature Materials* 3 (2004): 190–196.
5. E. Mathiowitz, J. S. Jacob, Y. S. Jong, G. P. Carino, D. E. Chickering, P. Chaturvedi, C. A. Santos, K. Vijayaraghavan, S. Montgomery, M. Bassett, and C. Morrell, "Biologically erodable microsphere as potential oral drug delivery system," *Nature* 386, no. 6623 (1997): 410–414.
6. M. E. Davis, S. H. Pun, N. C. Bellocq, T. M. Reineke, S. R. Popielarski, S. Mishra, and J. D. Heidel, "Self-assembling nucleic acid delivery vehicles via linear, water-soluble, cyclodextrin-containing polymers," *Current Medicinal Chemistry* 11 (2004): 1241–1253.
7. R. D. Bolskar, A. F. Benedetto, L. O. Husebo, R. E. Price, E. F. Jackson, S. Wallace, L. J. Wilson, and J. M. Alford, "First soluble M@C-60 derivatives provide enhanced access to metallofullerenes and permit in vivo evaluation of Gd@C-60[C(COOH)(2)](10) as a MRI contrast agent," *Journal of the American Chemical Society* 125 (2003): 5471–5478.
8. J. L. West and N. J. Halas, "Engineered nanomaterials for biophotonics applications: Improving sensing, imaging, and therapeutics," *Annual Review of Biomedical Engineering* 5 (2003): 285–297.
9. S. L. Tao and T. A. Desai, "Microfabricated drug delivery systems: From particles to pores," *Advanced Drug Delivery Reviews* 55 (2003): 315–328.
10. J. Santini, M. Cima, and R. Langer, "A Controlled-release Microchip," *Nature* 397 (1999): 335.
11. M. R. Prausnitz, S. Mitragotri, and R. Langer, "Current status and future potential of transdermal drug delivery," *Nature Reviews Drug Discovery* 3, no. 2 (2004): 115–124.
12. M. R. Prausnitz, "Microneedles for transdermal drug delivery," *Advanced Drug Delivery Reviews* 56, no. 5 (2004): 581–587.

Chapter 17

Bio-Nano-Information Fusion

Chih-Ming Ho, Dean Ho, and Dan Garcia

The solid-state microelectronics industry started in the late 1940s and has been the main driving force of the U.S. economy in the past few decades. In the 1980s, microelectromechanical systems (MEMS) technology demonstrated the capability of fabricating mechanical devices at a scale far smaller than those made of traditional machining processes. Fan, Tai, and Muller (1988) developed a micromotor that was on the order of 100 μm by using the MEMS fabrication process, which was derived from the integrated circuit (IC) industry.

Further reduction of the scale into the nanoscale has enabled the fusion between nanotechnology and biology. By working on the same scale as that of macro functional molecules, such as DNA and proteins, nanotechnology emerges as an opportunity to further build upon earlier technologies. An example is the nanomotor developed by Soong et al. (2000), which grew from the amalgamation of the mitochondrial ATPase and a nanofabricated metal rod. The progression of time has produced machines of diminishing size, leading the engineering sciences into progressively smaller dimensions. This movement has facilitated the expansion of the field of nanotechnology to explore and develop this new scientific frontier.

Technological innovations such as nanotechnology aim to enrich human life. However, nanotechnology presents a unique technical challenge in that a disparity of nine orders of magnitude separates the length scales of a human (a meter) and a nanometer. Ultimately, this goal needs to be achieved through the development of a definitive pathway that fuses existing technology with future technology to link the nanoscale with human life.

An extremely intelligent, complex, and adaptive system, the human body is managed by natural processes driven by molecules, such as DNA and proteins, that are on the order of a nanometer. The challenges in exploring the governing mechanisms across a wide span of length scales is best depicted by

P. W. Anderson in a paper published in *Science* (1972): "At each level of complexity entirely new properties appear, and the understanding of the new behaviors requires research which I think is as fundamental in its nature as any other." For example, a cell might fuse the genetic information contained in the DNA with nanoscale sensors and actuators to result in perhaps one of the most efficient and autonomous micron-scale "factories."

The richness of the science that spans a difference of three orders of magnitude in length scale, from the nanoscale realm to the macroscale realm, significantly exceeds our comprehension. An imperative task will be to address the question of how we will span the length scales of these nanoscale capabilities in a way that will eventually enable us to enrich human lives. These basic processes that occur at the molecular level have opened a world where the integration of individual components can eventually derive higher-order functionalities, or *emergent properties*. This leads us toward a compelling approach that fuses biotechnology, nanotechnology, and information science, which will enrich the development of revolutionary, application-specific technologies.

To drive the commercialization of nanotechnology, we need to further our understanding of the nanosciences. During the past decades, our understanding of nanoscale molecules and their functionalities has been significantly furthered by distinguished scientists in biology, physics, and chemistry. Nanotechnology, on the other hand, is still in its infancy, but it is the key to realizing a new industry. The establishment of the National Nanotechnology Initiative (NNI) has further increased the momentum of progress in these areas.

For this realization to occur, three goals must be achieved. First, core technologies must be developed to enable us to visualize, pick, place, manipulate, and characterize nanoparticles with a high degree of precision. Second, we will need to establish manufacturing technologies to enable systematic integration toward larger length scales with the higher information content embedded within the composite structures. The final step toward establishment of a nanotechnology industry will result from the capabilities displayed by emergent behavior at the nanoscale for increasing functionalities.

Instead of providing a comprehensive review, this chapter uses examples based primarily on our experience to illustrate the challenges and the potential associated with the impacting field of bio-nano-information fusion.

SEEING

Perhaps one of the most beneficial tools afforded by the process of realizing the full potential of nanotechnology will be the ability to visualize nanoscale molecular activity in real time. Continually advancing imaging methods have pushed the boundaries of nanoscale resolution.

Fluorescence microscopy greatly enhances the visualization of micro- and nanometer particles in biological specimens. For example, confocal microscopy augments the fluorescent labeling of everything from individual proteins to whole cells by offering several benefits over traditional fluorescent microscopy. Of particular note is that confocal microscopy reduces background fluorescence by using a filter that blocks out fluorescence located outside the focal point of the lens. Because it contains multiple excitation sources, confocal microscopy provides the further benefit of the simultaneous use and visualization of multiple fluorescent dyes. This allows, for example, the visualization of multiple neuronal cell attributes, as illustrated in Figure 17–1.

Atomic force microscopy (AFM) has evolved as a valuable method for evaluating surface topography down to the nanoscale, largely because of its incorporation of a flexible cantilever containing a nano-sized tip. Recently, AFM has been used to study nano-sized pores contained in cell membranes that facilitate the exchange of solutes and nutrients between their cytoplasm and their environment. For example, within its outer membrane the bacterium *Escherichia coli* contains channels called porins that open and close in response to changes in pH and transmembrane voltage. To study the conformational changes of the porins under the influence of pH and voltage, Muller and Engel (1999) employed atomic force microscopy. Upon varying the pH or voltage across the membrane, AFM confirmed conformational changes in the porin, showing a transition from a columnar structure to one consisting of a nano-sized hole, as shown in Figure 17–2.

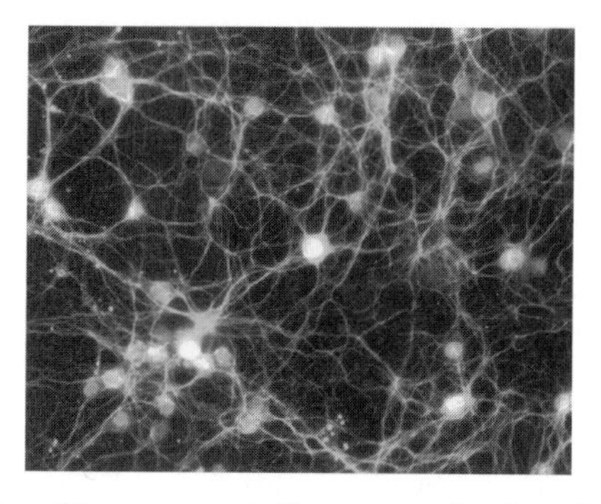

Figure 17–1 Ten-day-old mouse cerebellar neuronal culture labeled with neuronal marker beta III-tubulin and a marker for de novo DNA methyltransferase (Feng, Chang, Li, and Fan 2005).

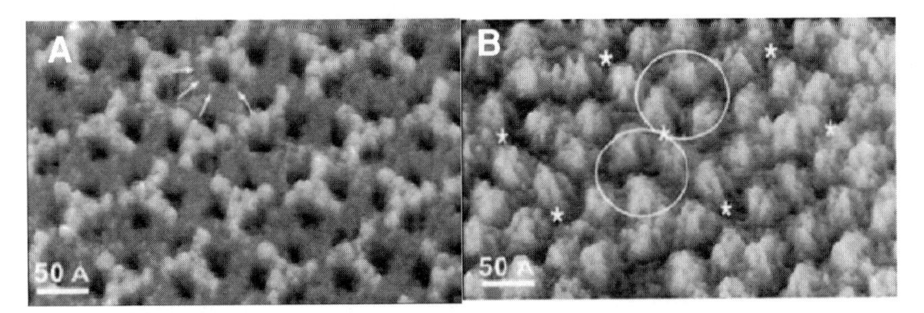

Figure 17–2 High-resolution AFM images of ion gradient and pH-dependent conformation changes of porin OmpF: (A) closed OmpF porin channel; (B) open OmpF porin channel (Müller et al. 1999).

Most recently, using materials that can amplify near field optical waves to visualize subjects far below the diffraction limit (Fang et al. 2003; Pendry 2001) will open a new territory for advancing nanotechnologies. The continued enhancement of the nanoscale visualization methods will enable discoveries of complex processes, including receptor-ligand interactions, DNA translocation across cellular membranes, and beyond. Such a capability will revolutionize our ability to directly observe reactions upon which the functionality of future nanotechnological devices will be based.

MANIPULATION

The exploitation of macro-functional molecules will inherently involve the application of their natural characteristics. Taking advantage of those interesting properties of nanoscale particles will require the successful manipulation of their position and properties. For example, a self-assembling biological system could be employed in a bottom-up fabrication scheme. The fusion of biology with nanotechnology will not only result in technological innovation but will also encourage future research into natural biological nanosystems.

Current technology harnesses the ability to deform individual molecules such as proteins or DNA. DNA is a nanoscale, long-chain macromolecule and behaves like a mass-spring system. Placing a DNA molecule inside a viscous flow stretches the DNA molecule from its coiled position at equilibrium by applying stress across the length of the molecule. To stretch DNA molecules, Wong et al. (2003) designed microfluidic channels with two buffer streams that converge to sandwich a middle stream containing the DNA solution, effectively extending the DNA molecule, as shown in Figure 17–3. Such a

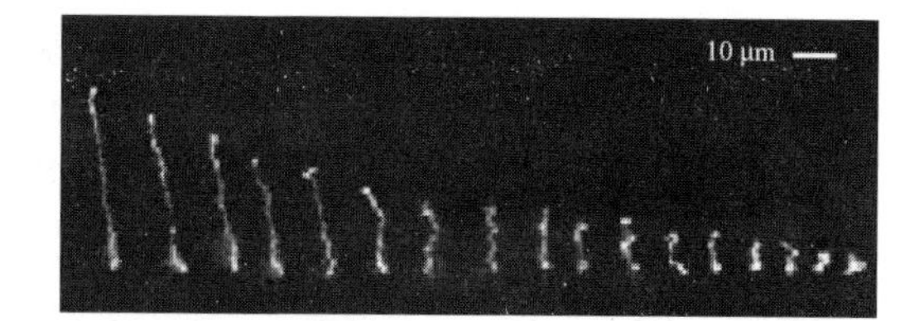

Figure 17–3 Relaxation of a DNA molecule after stretching is shown here in 2.5-second increments (Wong et al. 2003).

method proves advantageous in that the DNA extensional rates depend upon the flow rate of the outer buffer streams, thereby minimizing the effect of the flow from the centerline on the DNA. Furthermore, the low mixing among the streams enabled independent DNA control.

DNA/RNA Characterization

A precursor to the use of nanoparticles for practical, beneficial purposes will be the ability to characterize their properties with respect to composition, structure, and so on. Biomolecular characterization—with respect, for example, to proteins—has provided key information about three-dimensional structure as well as mechanistic behavior. Information gleaned from these studies serves as the foundation for current development of devices based purely on biomolecular function (Ho et al. 2004). Toward the realization of improving the human condition through nanotechnology, a full-scale characterization of DNA will elicit a broader understanding of the body of embedded information that governs this condition.

The mapping of the human genetic code as part of the Human Genome Project took ten years. Building on this foundation and developing a technology to quickly sequence an individual's DNA hold enormous potential for health maintenance and drug development. With specific respect to the interrogation of composition and structure, one pioneering methodology has been the use of nanoscale pores to characterize DNA molecules. For example, Meller et al. (2000) have monitored electrical conductance through an α-hemolysin channel from *Staphylococcus aureus* to yield important information regarding nucleotide composition and the effects of temperature on DNA transport.

Current results can discern transport events involving adenine polymers (poly dA_{100}) as well as cytosine polymers (poly dC_{100}) (Figure 17–4). These studies enable investigation of several characteristics of the analyzed molecule, from chain length to its exact composition and structure.

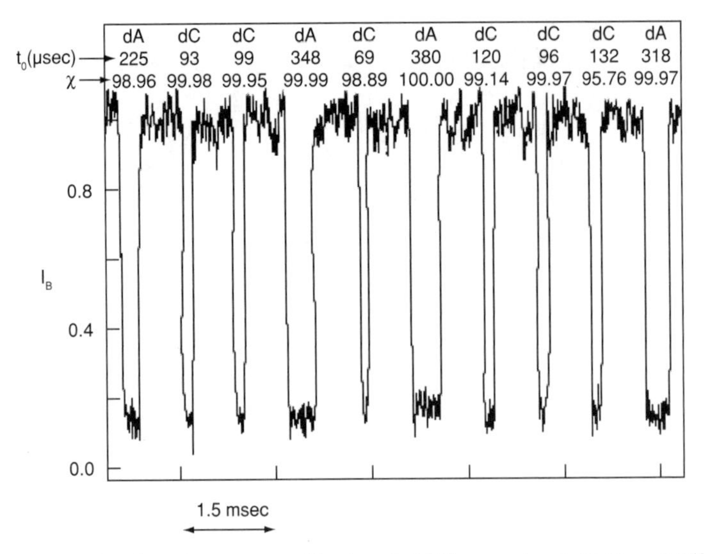

Figure 17–4 Translocation events of both poly(dA)$_{100}$ and poly(dC)$_{100}$. Differentiation between dA and dC translocation was determined using t_D provided in μs^{10} (Meller et al. 2000).

The use of membrane proteins to examine DNA strands possessed certain limitations, including the range of conditions in which characterizations could be performed, because the α-hemolysin and lipid membranes possessed their own ranges of optimal conditions for preserved activity.

To build upon the groundwork that was established by protein- and lipid-based characterization, Chen et al. (2004) used Si_3N_4-based nanopores to serve as solid-state characterization systems. Not confined to the same set of limitations observed with a protein-based setup, the solid-state nanopores were able to successfully detect translocation, or movement, of DNA through the pore while possessing the versatility of varied pore diameters to decrease the chances of blockages to DNA translocation and so on. Furthermore, solid-state nanopores will withstand broader ranges of pH, temperature, and pressure, as well as voltages that would normally impair the protein/lipid assembly, thereby allowing the testing of a broader class of molecules in a wider range of environments.

CHARACTERIZATION

Atomic force microscopy initially evolved as a valuable method for the evaluation of surface topography. Recently, AFM has been frequently used as a tool for the characterization and manipulation of nanoscale particles. For example, the field of AFM lithography is used to transfer nanoscale patterns onto photoresist, a UV or chemically reactive polymer commonly used with microfabrication. The unparalleled resolution achieved using AFM-based techniques makes it an ideal tool for nanolithography.

Furthermore, AFM has also been applied in studies involving biomolecules, such as membrane and motor proteins. Possessing the capabilities for intimate exploration of these biological systems, atomic force microscopy has been used for protein-folding measurements, as was done by Rief et al. (1997) with titin immunoglobulin domains. Understanding protein-folding mechanisms is important because it underlies the use of protein engineering, protein-based device engineering, and even potential applications in using the characterization of a person's protein-folding properties as a means of health monitoring.

The use of atomic force microscopy is a prime example that demonstrates how existing technologies have been applied to the understanding of a requisite component of future nanotechnological systems.

INTEGRATION

The human benefit of advances in nanotechnology will stem from the transition from handling single nanoparticles to realizing large-scale manufacturing capabilities. Spanning from the nanoscale to the macroscale is a necessary process in the transition of the nanosciences into industry. Artificial fabrication technologies are usually considered top-down, meaning that they involve manipulations at the macroscale, such as cutting or etching bulk materials. Nature, on the other hand, relies on bottom-up fabrication, involving molecule-by-molecule assembly to create a predesigned subject. Thus, the goal of artificial bottom-up fabrication involves the directed and orderly integration of molecules at the nanoscale to form a macroscale device.

An example of bottom-up fabrication is the self-assembling monolayer (SAM) used by chemists (Ulman 1991, 1995), such as the attachment of nano-sized motor molecules to cantilever beams to form actuators. Rotaxane, a motor molecule shown in Figure 17–5, has two separate recognition sites, each associated with a different energy potential. Placing the rotaxane molecule in an oxidant oxidizes one site, thereby raising the energy level associated with the site (Collier 1999). As a consequence, the energy profile of the rotaxane molecule is altered so that the ring prefers association with the other recognition site, thereby causing the ring to move. Applying a reductant will

Figure 17–5 Graphical representation of a rotaxane molecule. Alternating oxidation and reduction of the molecule results in movement of the ring structures (Collier et al. 1999).

return the energy profile to its original state, forcing the ring back to its original position. In this way, a molecular motor is created. This is an example of how to use artificial self-assembly to integrate atoms into a molecular system.

The rotaxane molecule was further modified to create a molecule containing two rings and four recognition sites. Anchoring the two rings to a gold surface creates a molecular muscle that, when placed in an oxidant, causes the two rings to move to the center of the molecule and bend the beam, a movement that can be detected by a laser beam. An array of molecular muscles attached to cantilever beams was created, as shown in Figure 17–6. Alternating the application of oxidant and reductant causes the beam to bend upward and return to its original position. This experiment illustrates how we com-

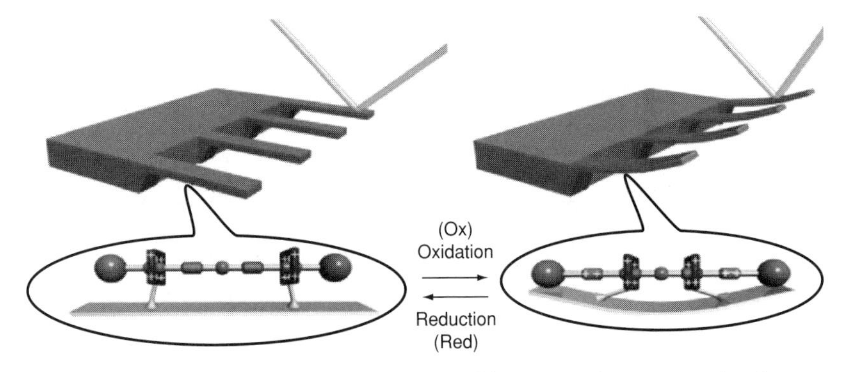

Figure 17–6 A microcantilever beam activated by nanoscale motor molecules (Huang et al. 2004).

bine the top-down and the bottom-up fabrication techniques to form an integrated micromechanical system from the nanoscale integrated system.

EMERGENT SYSTEMS

A human cell serves as a culmination of what nature has taken millions of years to evolve: an autonomously responsive system of sensors and actuators that operates based on commands from an embedded and distributed intelligence. It is a self-regulating, self-governing unit, with its nucleus serving as its central information processor, and hence it represents a system based on the fusion of a number of factors. The transfer and transduction of information using various signal pathways found in cells serve to process and apportion this information to induce a concerted action. For example, a chemical signal results in a sensory response that elicits mechanical movement or actuation from the cytoskeletal network—all characteristic behavior in chemotaxis, or chemical-induced cellular movement.

As a composite system, a cell exemplifies the concept of emergence, where inputs result in coordinated feedback. The example of a neutrophil hunting down and encircling the *Staphylococcus aureus* bacterium (Figure 17–7) demonstrates concerted, self-determining behavior. The bacterium emits a chemical gradient that is sensed by a cell. The cell, which is able to follow a complex path of the chemical, moves toward the bacterium and eventually

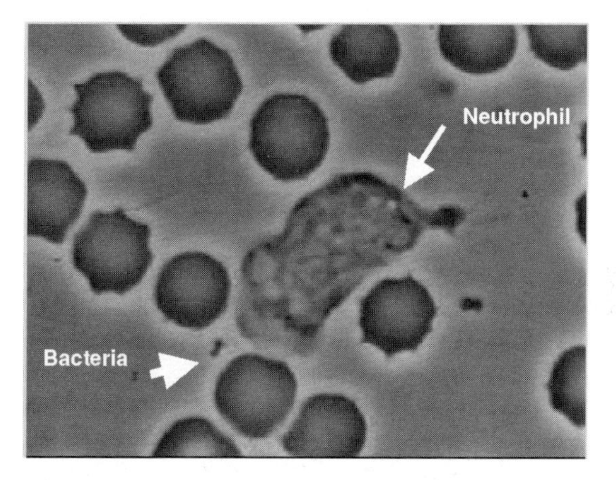

Figure 17–7 A neutrophil is observed chasing a *Staphylococcus aureus* bacterium. The bacterium emits a chemical that is in turn sensed by the neutrophil, which then coordinates an autonomous actuation directed toward the phagocytosis of the bacterium (Rogers et al. 1950).

surrounds the bacterium to envelope, or phagocytoses, it. In this process, the chemical is sensed by the chemical sensors inside the neutrophil and is processed by the signal pathways. Eventually, the neutrophil acts as a whole to capture the bacterium.

A key element of manufacturing large-scale molecular systems will be the derivation of emergence, or true mimicry, toward applications in energy production, nanoscale medicine, and so on. Continued progress in nanotechnological development will result in promising approaches whereby the input of stimuli (such as light, reactive chemicals, and so on) induces systemic behavior not previously present to produce emergent behavior.

Showing particular promise are micrometer-sized photonic crystals of porous silicon presented by Link and Sailor (2003). The optical properties of these crystals change upon the absorption of chemicals, the fabrication of which is illustrated in Figure 17–8.

The "smart dust" particles are composed of two sides: a green, "water-fearing" (hydrophobic) side, and a red, "water-loving" (hydrophilic) side. Furthermore, because of their amphiphilic nature, the smart dust particles will orient themselves spontaneously at a water surface to form a monolayer so that the hydrophilic side faces the water and the hydrophobic side faces the air.

If porous silicon particles are incubated in water with a drop of dicholoromethane solvent, the smart dust particles will self-assemble and orient themselves around the drop of dichloromethane so that the hydrophilic side (red) faces the water and thc hydrophobic side faces the solvent drop, as shown in Figure 17–8. The individual particles aggregate together to form a large, macroscopic collection that *emerges* as a result of the particle self-assembly. As demonstrated by Link and Sailor, the smart dust particles prove

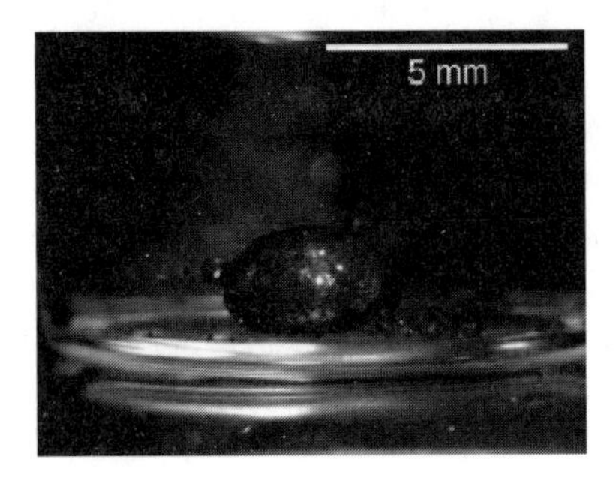

Figure 17–8 Surrounding of dichloromethane drop by self-orienting, self-assembling "smart dust" particles. This is accomplished through hydrophobic-hydrophobic interaction between the dichloromethane and the hydrophobic porous silicon (Link et al. 2003).

useful in the detection of a chemical such as dichloromethane. The modification of the particles with recognition elements may add further use to the particles by facilitating the detection and isolation of pathogenic organisms in food or water.

Beyond the practical application of this work to detection and similar uses, a compelling approach to deriving an intrinsic, higher-order behavior from the system has been established. This was accomplished through the addition of a new condition to the solution, represented by the drop of dichloromethane.

We are currently at a burgeoning stage with respect to the development of emergent behavior in artificial systems, and continued efforts will seek to embed increased quantities of information in artificial systems to derive even more complex higher-order functionality, such as usable energy or other coordinated activity. The push toward true fusion will inevitably arrive at the achievement of true mimicry. This will be an essential precursor to realizing the human benefit of nanotechnology.

CONCLUSIONS

This work has outlined the requisite strategies for realizing the ultimate goal of nanotechnology: to benefit the human condition. Fundamental studies in the nanosciences have provided the building blocks on which nanotechnology will drive the creation of novel devices with applications in energetics, electronics, materials, medicine, and beyond. These systems will address the differences between being integration- or fusion-based by possessing embedded intelligence through a series of nanoscale sensors and actuators. The example of coordinated smart dust activity provides a promising demonstration of basic emergent behavior. Future work will seek to dramatically increase the amounts of information contained within fabricated systems to provide progressively advanced outputs in response to various impulses. For example, by inputting a specific stimulus (such as sunlight) into these devices, an autonomous reactivity (peptide-driven energy transduction) will ultimately produce a usable output (electricity).

Bridging the elements of nanotechnology with beneficial larger-scale systems will parallel current and continued advancements in methodologies with which nanotechnology will develop. These will include the handling abilities to enable us to see and manipulate nanoparticles with the goal of characterizing them. Manufacturing strategies will transition single-molecule studies toward large-scale fabrication of systems that possess increasing information content. Finally, the realization of systems based on the fusion of biology, nanotechnology, and informatics will result in truly emergent, or biomimetic, systems that combine sensors and actuators that respond to specific stimuli.

Although the nanotechnology industry is in a nascent stage, rapid advancements and a streamlined road map of progress ensure that the future is quite promising. The realization of this industry's potential will have revolutionary and compelling impacts upon humankind. For example, the fruition of rapid DNA screening and diagnosis modalities will open the gateway to designing custom therapeutics tailored to individuals based on their genetic makeup. An achievement of this magnitude would serve dual roles. First, nanotechnology would be cemented as the visionary industry for the next millennium. Second, the true benefits for humankind enabled by the maturation of this technology will have been realized.

However, to reach this point in the road map, we must address in depth several key areas previously outlined. Through the use of emerging technologies and methodologies for discovery—such as the use of superlens material for direct imagery of nanoscale processes or the enhancements of single-molecule manipulation abilities—we will achieve an unprecedented, more complex level of control of biological processes. This, in turn, will give us a deeper understanding of how these biomolecules and their respective activities contribute to a global functionality (such as the systemic performance of the human body) to create an emergent behavior in nature. In this way, nanotechnology will then be poised to reproduce this behavior to combat disease, to produce advanced energy sources, and to support other advancements that will redefine the way we livc.

ACKNOWLEDGMENTS

The authors would like to gratefully acknowledge The Institute for Cell Mimetic Space Exploration (CMISE, a NASA URETI), the National Institutes of Health (NIH), the National Science Foundation (NSF), and the Defense Advanced Research Projects Agency (DARPA) for the support of this work.

REFERENCES

Chen, P., Gu, J., Brandin, E., Kim, Y., Wang, Q., and Branton, D., "Probing single DNA molecules transport using fabricated nanopores," *Nano. Lett.* 4 (2004): 2293.

Collier, C. P., Wong, E. W., Belohradsky, M., Raymo, F. M., Stoddart, J. F., Kuekes, P. J., Williams, R. S., and Heath, J. R., "Electronically Configurable Molecular-Based Logic Gates," *Science* 285 (1999): 391.

Fan, L. S., Tai, Y. C., and Muller, R. S., "IC-processed Electrostatic Micromotors," *Technical Digest, IEDM* (1988b): 666.

Fang, N., and Zhang, X., "Imaging properties of a metamaterial superlens," *Applied Physics Letters* 82, no. 2 (2003): 161.

Feng, J., Chang, H., Li, E., and Fan, G., "Dynamic Expression of De Novo DNA Methyltransferases Dnmta3a and Dnmt3b in the Central Nervous System," *Journal of Neuroscience Research* 79, no. 6 (2005): 734.

Ho, D., Chu, B., Lee, H., and Montemagno, C. D., "Protein-driven energy transduction across polymeric biomembranes," *Nanotechnology* 15 (2004): 1084.

Huang, T. J., Brough, B., Ho, C. -M., Liu, Y., Flood, A. H., Bonvallet, P. A., Tseng, H. -R., Stoddart, J. F., Baller, M., and Magonov, S., "A Nanomechanical Device Based on Linear Molecular Motors," *Applied Physics Letters* (November 29, 2004).

Huang, T. J., Tseng, H. -R., Sha, L., Lu, W., Brough, B., Flood, A. H., Yu, B. -D., Celestre, P. C., Chang, J. P., Stoddart, J. F., and Ho, C. -M., "Mechanical Shuttling of Linear Motor-Molecules in Condensed Phases on Solid Substrates," *Nano Letters*, in press.

Link, J. R., and Sailor, M. J., "Smart dust: Self-assembling, self-orienting photonic crystals of porous Si," *Proceedings of the National Academy of Sciences* 100, no. 19 (2003): 10607.

Meller, A., Nivon, L., Brandin, E., Golovchenko, J., and Branton, D., "Rapid nanopore discrimination between single polynucleotide molecules," *Proc. Nat. Acad. Chi.-USA* 97 (2000): 1079.

Müller, D. J., and Engel, A., "Voltage and pH-induced Channel Closure of Porin OmpF Visualized by Atomic Microscopy," *Journal of Molecular Biology* 285 (1999): 1347.

Pendry, J., "New electromagnetic materials emphasise the negative," *Physics World* (2001).

Rief, M., Gautel, M., Oesterhelt, F., Fernandez, J., and Gaub, H., "Reversible unfolding of individual titin immunoglobulin domains by AFM," *Science* 276 (1997): 1109.

Rogers, D., "Neutrophil crawling," video, 1950, http://expmed.bwh.harvard.edu/projects/motility/neutrophil.html.

Soong, R. K., Bachand, G. D., Neves, H. P., Olkhovets, A. G., Craighead, H. G., and Montemagno, C. D., "Powering an Inorganic Nanodevice with a Biomolecular Motor," *Science* 290 (2000): 1555.

Ulman, A., *An Introduction to Ultrathin Organic Films: From Langmuir-Blodgett to Self-Assembly* (Boston: Academic Press, 1991).

Ulman, A., *Characterization of Organic Thin Films* (Boston: Butterworth-Heinemann, 1995).

Wong, P. K., Lee, Y. K., and Ho, C. M., "Deformation of DNA molecules by hydrodynamic focusing," *J. Fluid. Mech.* 497 (2003): 55.

SECTION FOUR

Convergence and Integration

Chapter 18

Convergence and Integration

Mihail C. Roco

Unifying science based on the material unity of nature at the nanoscale provides a new foundation for knowledge, innovation, and integration of technology. Revolutionary and synergistic advances at the interfaces between previously separated fields of science, engineering, and other areas of relevance are ready to create nano-bio-info-cogno (NBIC) transforming tools, products, and services. Developments in systems approaches, mathematics, and computation in conjunction with NBIC allow us to understand the natural world and scientific research as closely coupled, complex, hierarchical entities.

At this unique moment of scientific and technical achievement, it becomes possible to improve human performance at individual and group levels, as well as to develop suitable revolutionary products. These are primary goals for converging new technologies. Most nanotechnology applications are realized at the confluence of nanotechnology with other technologies. NBIC addresses long-term advances in key areas of human activity, including working, learning, aging, group interaction, organizations, and overall human development (Roco and Bainbridge 2003). Fundamentally new tools, technologies, and products will be integrated into individual and social human architectures. This chapter outlines research and education trends and discusses the potential for the development of revolutionary products and services.

FRAMEWORK FOR SCIENCE AND TECHNOLOGY DEVELOPMENT

In the next ten or twenty years, concentrated efforts from a number of disciplines are likely to bring greater unity to science—a reflection of the unity of the natural world. This foundation will lead to the synergistic combination of four major provinces of science and technology, each of which is currently progressing at a rapid rate in connection with the others: (1) nanoscience and nanotechnology; (2) biotechnology and biomedicine, including genetic engineering; (3) information technology, including advanced computing and communications; (4) cognitive sciences, including cognitive neuroscience and systems approach concepts. Their synergistic integration from the nanoscale is expected to bring significant advances. Collectively, the convergence of these scientific and technological domains is here referred to as NBIC, shorthand for the various fields and subfields encompassed by the *nano-*, *bio-*, *info-*, and *cogno-* prefixes. NBIC convergence is using and interacting with the broad base of more established technologies.

With proper attention to ethical issues and societal needs, these converging technologies could lead to a tremendous improvement in human abilities, societal outcomes, U.S. productivity, and the quality of life. Six areas of relevance for human performance have been identified (Roco and Bainbridge 2003; Radnor and Strauss 2004) as most important: human cognition and communication, health and physical capabilities, group and societal outcomes (including new products and services), national security, science and education, and business and organizations.

The integration of NBIC tools is expected to lead to fundamentally new products and services, such as entirely new categories of materials, devices, and systems for use in manufacturing, construction, transportation, medicine, emerging technologies, and scientific research. Fundamental research will occur at the confluence of physics, chemistry, biology, mathematics, and engineering. Nanotechnology, biotechnology, and information technology will play an essential role in the research, design, and production of these new products and services.

Industries increasingly will use biological processes in manufacturing. Examples are pharmaceutical genomics; neuromorphic technology; regenerative medicine; biochips having complex functions; molecular systems having multiscale architectures; electronic devices having three-dimensional, hierarchical architectures; software for realistic multiphenomena and multiscale simulations; processes and systems phenomena from the basic principles at the nanoscale; new flight vehicles; and quantitative studies using large databases in social sciences. Cognitive sciences will provide better ways to design and use the new manufacturing processes, products, and services, as well as lead to new kinds of organizations.

Within the U.S. Government, the National Science Foundation (NSF), the National Aeronautics and Space Administration (NASA), the Environmental Protection Agency (EPA), the Department of Defense (DoD), and the

Department of Energy (DOE) already have several seed R&D projects in the area of converging technologies. These projects are based on unifying science and education, creating infrastructure for research at the confluence of two or more NBIC domains, developing neuromorphic engineering, improving human performance, advancing "learning how to learn," and preparing for the societal implications of converging technologies. Ethical and other societal implications must be addressed from the beginning of any major program. Industry involvement is evident in seed projects and in strategic R&D plans of several companies. User and civic group involvement is essential if we are to take advantage of the technology sooner and develop a complete picture of societal implications.

We need a systematic, deliberative, and responsible approach. This chapter briefly outlines the key areas of relevance to converging NBIC technologies, several trends, and current NBIC outcomes.

The Opportunity

Advancing a coherent approach for converging technologies with a focus on human potential, increased productivity, and revolutionary products and services is timely for five main reasons:

1. Accelerated human progress has become possible at the individual and collective levels. We have arrived at the moment when we can measure signals from, and interact with, human cells and the nervous system. We have also begun to replace or regenerate body parts as well as build machines and other products that are suitable for direct interaction with human tissue and the nervous system.
2. Unifying science based on the material unity at the nanoscale, and knowledge integration from that scale, will provide a new foundation for technological innovation and the development of humanities (philosophy, the arts, and so on). This implies the need to understand biosystems and changes in matter at their roots. Nanoscale concepts have been rejuvenated by new tools for measurement, control, and manipulation. The systems approach has been powered by advances in hierarchical architecture, mathematics, and information technology. The natural laws of interdependence were recognized by many ancient civilizations. However, because they lacked a coherent understanding of connections, the concepts were reflected only philosophically. More than five centuries ago, the leaders of the Renaissance saw "unity in nature," but this concept was followed by disciplinary specialization because of limited integrative knowledge. Only now can we begin to connect and

integrate various disciplines and provide cause-and-effect explanations from the nanoscale to the macroscale, moving beyond the Renaissance ideal.

3. The accelerated rate and scale of changes in key technologies suggest the need for a fresh approach. The context for the development of new products and processes evolves rapidly. Several paradigm changes are expected in each of the four NBIC domains:
   - Nanotechnology is expected to move from its current focus on scientific discovery toward a focus on systematic design methods and technological innovation, leading to the development of manufacturing methods for mass production.
   - Biotechnology will move toward molecular medicine, nanosystems approaches, and pharmaceutical genomics, and biomaterials will be integrated at an increased rate in industrial products.
   - Information technology, in the quest for smallness and speed, will be enhanced by a focus on new architectures, three-dimensional design, functionality, and integration with application-driven developments in areas such as biosystems and knowledge-based technologies. A special opportunity will be created by the ability to analyze large, complex, hierarchical systems.
   - Cognitive sciences will focus on explaining the brain, the mind, and human behavior based on an understanding of physical, chemical, and biological processes at the level of the neuron, and from a systems approach. Increased attention will be given to human–technology coevolution. Technologies based on large, hierarchical, dynamic, complex systems will address new opportunities such as quantitative social sciences.
4. Because of the significant impact of NBIC convergence, there is a need for anticipation ("learning before doing"), deliberate choices (for logical and democratic decisions), and precautionary measures (for unexpected consequences).
5. Science and technology are increasingly recognized as the main sources of overall human development (UNPD 2001; Greenspan 2001).

A new infrastructure based on the four NBIC research and development platforms will be necessary to create the products listed here. Ideally, this infrastructure must be available anywhere, on short notice, to any industry and all those interested. A broader range of R&D issues will be investigated. For example, R&D challenges in nanoscale engineering are as follows: three-dimensional architectures that incorporate materials, devices, systems, space, and time; directed assembling, patterning, and templating of heterogeneous nanosystems, multiphenomena, and multiscale design; integration of nanoscale elements into larger scales; the creation and use of intermediary standard components; thermal and chemical stability of nanostructures; operational and environmental safety; and reliability and reproductivity at the nanoscale.

Criteria for Progress

To measure and better evaluate performance, it will be necessary to adopt new socioeconomic indices; one example is a modified GNP to incorporate changes in the human condition, to reflect the impact on the environment, to prepare the infrastructure (including education and training), and to measure other annual accumulations having societal implications. New holistic criteria may be considered for evaluating progress, such as reducing the entropy of a system—that is, less energy dissipation per computation and transmission of information; less material, energy, water, and pollution in nanotechnology; and less change or degradation in biotechnology. New indices are necessary for recognizing accumulations in infrastructure, a better-educated workforce, more-productive tools, and knowledge creation during a time interval.

OUTLOOK FOR INDUSTRY

Ongoing R&D programs in nanotechnology promise to increase the efficiency of traditional industries and to bring radically new applications, for the following reasons:

- Numerous newly designed, advanced materials and manufacturing processes will be built by 2015 using control at the nanoscale level in at least one key component. Silicon transistors will reach dimensions of 10nm or smaller and will be integrated with molecular or other kinds of nanoscale systems.
- Suffering from chronic illnesses has already been reduced through advances in technology. It is conceivable that by 2015, techniques for detecting and treating tumors in their first year of occurrence could be developed that might greatly mitigate suffering and death from cancer.
- The convergence of science and engineering at the nanoscale level will establish a mainstream pattern for applying and integrating nanotechnology with biology, electronics, medicine, learning, and other fields. The science and engineering of nanobiosystems will become essential to human health care and biotechnology. Life-cycle sustainability and biocompatibility will be pursued in the development of new products.

- Knowledge development and education will originate from the nanoscale instead of the microscale level. A new education paradigm based, not on disciplines, but on unity of nature and integration of education and research, will be tested for educational grades K–16.
- Nanotechnology businesses and organizations will be restructured with an eye toward integration with other technologies, distributed production, continuing education, and the formation of consortia of complementary activities. Traditional and emerging technologies will be equally affected.

Nanotechnology's capabilities for systematic control and manufacture at the nanoscale level will evolve in four overlapping generations of new products. Each generation is marked here by the creation of the first commercial prototypes using systematic control of the respective phenomena and manufacturing processing:

- The first generation of products emerged in 2001 in the form of passive nanostructures used to tailor macroscale properties and functions: nanostructured coatings, the dispersion of nanoparticles, and bulk materials such as nanostructured metals, polymers, and ceramics.
- Second-generation products—active nanostructures for mechanical, electronic, magnetic, photonic, biological, and other effects, integrated into microscale devices and systems—will begin to emerge several years after the first generation. Examples might include new transistors, components beyond CMOS amplifiers, targeted drugs and chemicals, actuators, artificial muscles, and adaptive structures.
- Within five to ten years of the second generation, a third can be expected. These products will include nanosystems with 3-D features, using various syntheses and assembling techniques (such as hierarchical, self-organizing bioassembling robotics with emerging behavior), as well as evolutionary approaches. Key challenges include networking at the nanoscale and developing hierarchical architectures. Research focus will shift toward engineering heterogeneous nanostructures and supramolecular systems, some with emerging or evolutionary behaviors. This includes directed multiscale, self-assembling, artificial tissues and sensorial systems; quantum interactions within nanoscale systems; processing of information using photons or electron spin; assemblies of nanoscale electromechanical systems (NEMS); and the convergence of nano-, bio-, info-, and cognitive platforms integrated from the nanoscale level and up.

- Within five to ten years of the third generation, the fourth generation can be expected to bring heterogeneous molecular nanosystems, where each molecule in the nanosystem has a specific structure and plays a unique role. Molecules will be used as devices. From their engineered structures and architectures will emerge fundamentally new functions. The design of new atomic and molecular assemblies is expected to increase in importance. This includes macromolecules "by design"; nanoscale machines; directed and multiscale self-assembly that exploits quantum control; and nanosystem biology for health care, the human-machine interface at the tissue and nervous-system level, and the convergence of nano-bio-info-cognitive domains.

Although expectations for nanotechnology may be overestimated in the short term, the long-term implications for health care, productivity, and the environment appear to be underestimated.

CONCLUDING REMARKS

The natural laws of interdependence were recognized by many ancient civilizations, and the idealized concept of unity in nature was promulgated during the Renaissance more than five centuries ago. However, only now have we begun to connect and integrate various disciplines based on recent advancements in knowledge and to provide cause-and-effect explanations from the nanoscale to the macroscale, thereby moving beyond the Renaissance ideal. With the proper consideration of ethical aspects and societal needs, paradigm changes and the synergism of the four components of nano-bio-info-cogno are expected to bring revolutionary changes in technology, the economy, and society, as well as to increase human potential.

ACKNOWLEDGMENTS

Opinions expressed here are those of the author and do not necessarily represent the position of the National Science and Technology Council or the National Science Foundation.

REFERENCES

Greenspan, A. 1999. Public statement of the Joint Economic Committee of the U.S. Federal Reserve, on June 14, 1999, Washington, DC.

Radnor, M. & J. Strauss, Eds. 2004. Commercialization and Managing the Converging New Technologies. Report sponsored by the National Science Foundation, Arlington, VA.

Roco M.C. & W.S. Bainbridge, Eds. 2003. Converging Technologies for Improving Human Performance. NSF-DOC Report. Boston: Kluwer.

Roco, M.C. & W.S. Bainbridge, Eds. 2004. Annals of the New York Academy of Sciences 1013: 1–259.

United Nations Development Programme. 2001. Human Development Report: New York: Oxford University Press.

Chapter 19

Ethical Considerations in the Advance of Nanotechnology

William Sims Bainbridge

Ethics is important in industry for a variety of reasons. A well-established professional ethics is a necessary basis for cooperation between companies, government agencies, and other organizations. The public needs to have confidence in the ethical conduct of a corporation's business, or it may be suspicious of the corporation's products and services in the marketplace. Without widespread confidence from academia and nonprofit organizations, a corporation may find itself the target of negative activism and political agitation.

Ethical principles are among the most significant means by which a company builds morale and support from its own employees, including the technical and scientific staff who are relied upon for profitable innovation. Stockholders and potential investors know that the reputation of a company is among its most valuable assets. Finally, without a well-articulated and seriously implemented set of ethical principles, the management of a corporation cannot take full pride in its own accomplishments nor be confident it is making the right decisions for the well-being of the company and of society at large.

Ethical questions of many kinds may arise in the general area of research, manufacturing, and application of nanotechnologies. For example:

- Who is responsible for preventing and dealing with possible harm to health or the environment?
- How can intellectual property rights be defined and defended when the boundary between nanoscience and nanotechnology is so ambiguous?
- How can the public's right to know about technologies that may affect their lives be balanced against an innovator's right to protect trade secrets and against the cost of collecting and disseminating correct information?

- What priority should be placed on developing new products and techniques that will be highly beneficial for some people, versus defending the interests of individuals and groups that may not benefit from them?

Although this chapter draws upon a number of sources, it especially reflects the deliberations that led to two major reports on the societal implications of nanoscience and nanotechnology (Roco and Bainbridge 2001, 2005).

THE NATURE OF ETHICS

It would be convenient if there were a complete consensus both about the rules for ethical behavior and about the best scholarly and scientific approach for analyzing ethical problems. Unfortunately, beyond a few very simple principles that are more a reflection of Western democratic institutions than of deep philosophical thought, there is no consensus. One reason is that several distinct societal institutions and academic disciplines make it their business to address ethical issues, and they do not agree entirely about fundamental principles. Another reason is that competing interests and ideologies take opposing positions on key issues.

The traditional source of ethical principles is religious doctrine (Stark and Bainbridge 1987, 1996). Western religions are often seen to be supportive of technological progress, in part because they often conceptualize the world as an environment prepared by God for human benefit, and they may go so far as to suggest that God has given humanity dominion over nature. However, specific technologies may run afoul of one or another traditional doctrine, as is the case with human reproductive cloning in the judgment of many theologians and other believers (Bainbridge 2003). A good continuing source of theologically based analysis of the ethics of science and technology is the journal *Zygon*.

In modern societies, with traditions of the separation of church and state, there has been hope that firm ethical principles could be established on a secular basis, perhaps by academic philosophers. Philosophy is the academic discipline most dedicated to asking penetrating questions and doubting widespread assumptions. As such, it is an excellent part of a well-rounded liberal education, but it may be poorly prepared to give definitive answers to the questions it raises. Nonetheless, academic training in the philosophy of ethics is a valuable part of preparation for decision makers or their advisers, because it helps them see important factors that may not be obvious—not only potential ethical problems but also solutions to them.

Participants at the conference *Societal Implications of Nanoscience and Nanotechnology II: Maximizing Human Benefit* (Roco and Bainbridge 2005) were divided on the issue of the contribution professional ethicists can make. Philosophers of ethics analyzed issues relating to equity and the quality of life, contributing insights that seemed valuable. However, other participants doubted whether academic ethicists possess the knowledge about nanotechnology or reliable techniques of philosophical analysis that could give their views greater weight than those of interested amateurs among the general public, or could arrive at judgments that were not simply political in nature.

A fundamental debate in philosophy concerns the extent to which good and evil can be objectively defined or instead are a matter of a judgment made by human beings from their particular standpoints and interests (Kant 1787; Moore 1951; Rawls 1971). The extreme view, expressed most famously by German philosopher Friedrich Nietzsche (1886–1887), is that morality is a scam practiced by dominant people to gain power over submissive people, and unfortunately many people today behave as if they believed this theory. A more manageable position is that morality is negotiated between people in order to enhance cooperation and to serve their mutual interests and their enlightened self-interest (Homans 1974; Gauthier 1986).

By means of traditional political procedures, government agencies establish rules in such areas as environmental protection and workplace safety; legislatures write laws regulating industry and prohibiting actions that are deemed to be dangerous; courts apply and interpret the law in particular cases brought to them by plaintiffs who have their own interests and often intense feelings. At times, the actions of social movements trump those of these other political actors, especially when a movement is able to set the terms of debate about a particular issue. Some individual promoters of nanotechnology have spread unreasonable views about its power for good or evil, and there is a danger that their views are defining the issues. In the modern world, it is impossible to escape the fact that ethical issues are always political issues.

Both sociology and anthropology study the factors and processes that establish, sustain, and challenge morality. Thus, these disciplines have much wisdom to offer concerning the dynamics of ethical debates. Sociologists and anthropologists tend personally to sympathize with powerless groups in society, and to be somewhat suspicious of the corporations and government agencies that generate new technologies. Perhaps ironically, this bias illustrates one of the most valuable insights they have to offer: Ethical principles are not objective or uniform across humanity but instead are rooted in specific social groups. The sociologist notes that ethical principles reflect the material interests of the people who promote them, whereas the anthropologist celebrates the cultural relativism of varying moralities across radically different societies.

In practical terms, this perspective can be very useful in dealing with a concrete ethical dispute. On the one hand, it suggests that it may not always be possible to achieve an ethical consensus and that it can be especially difficult to bring opponents around to one's own way of thinking. Although these observations seem pessimistic, they can save wasted effort in trying to achieve an agreement that is actually out of reach.

On the other hand, awareness that other people have their own distinct needs and values can help one learn to accommodate those interests. In addition, awareness that others may have really different cultural beliefs and values can facilitate honest bargaining with them. Understanding need not mean surrender, but it can help frame issues in ways that have the best chance of finding mutually satisfactory compromises.

ETHICS OF INDIVIDUAL BEHAVIOR

Sociology, social psychology, and criminology suggest insights that can help an organization encourage individuals to adhere to a code of ethical behavior (Bainbridge 1997, 2004). For example, an industrial corporation may have established very enlightened standards of behavior with respect to environmental pollution, workplace hazards, and intellectual property rights. But this may not do much good if social disorganization, pressures to overachieve, and internal conflict impel many members of the organization to violate these rules. Here are three suggestions that can be applied to work with nanotechnology but that also apply very generally:

1. Do not consistently leave any members of the group out of the communication channels or the network of social relationships. An extensive body of scientific research has found that individuals whose bond to society is weak or broken are no longer under social control and thus are especially free to violate rules.
2. Do not allow any individual to be under such extreme pressures to perform that he or she will feel forced to take unethical shortcuts. Job performance pressures can motivate some individuals to innovate in valuable ways, but when people's self-esteem and economic well-being depend on accomplishing things that are beyond their abilities, they are apt to take shortcuts or to engage in deception.

3. Do not allow any subgroups or divisions of the organization to become so autonomous that they develop a deviant subculture. Group cohesion and solidarity are valuable assets, as long as they reinforce dedication to the legitimate goals of the organization. Too much solidarity within a division, however, can cause the group to secede socially and psychologically from the larger organization, and excessive cohesion encourages *groupthink*, which is a harmful tendency to make decisions without considering valid criticism or the wider ethical context.

NANO-SPECIFIC ISSUES

Nanotechnology is primarily a multiplier for other technologies, providing enhanced performance, reliability, or economy. However, with respect to health and environmental issues, nanotechnology may pose some unique hazards (Royal Academy of Engineering 2003; Meridian Institute 2004). In many cases, the chemical, electrical, and mechanical properties of nanoscale particles are measurably different from those of bulk samples of the same substances.

Thus, it is essential to develop good methods for characterizing nanomaterials (for example, to determine the size distribution of particles) and good theories that would allow one to estimate the probable risk, if any, of a particular nanomaterial on the basis of a growing body of knowledge about the category to which it belongs.

It will be impossible to subject all distinguishable kinds of nanomaterials to extensive testing. Thus, direct empirical research is required on the impacts of a range of representative nanomaterials, research that then can be the basis for comparison of similar materials that are not tested so intensively. Risk assessment for nanomaterials covers four related steps—hazard identification, hazard characterization, exposure assessment, and risk calculation—which then prepare the way for risk management (Luther 2004: 43).

Hazards must be assessed realistically in terms of the life cycle of the technology: Research and development hazards may differ from those in manufacturing, in the use of the product, and in disposal when the product is no longer serviceable. Suppose research does find that carbon nanotubes cause health problems when injected into animals in substantial doses. Suppose also that carbon nanotubes become the key component in future manufacture of nanoelectronics, comparable to computer chips. During the period of their actual use, and presumably after disposal as well, a very small physical mass of nanotubes would be sealed safely inside the electronic components of a device, posing no realistic threat to anyone. But there might possibly be a period of measurable hazard at the point of manufacture, when the nanotubes were being generated and assembled into components. This, then, would be an issue for workplace ethics—to be negotiated by management and labor in the context of government regulations—but would not raise ethical questions concerning the general public.

Hazards may be higher in cases of accidents or misuse. For example, nanoscale-engineered substances designed to hold hydrogen in the fuel tanks of future hydrogen-powered automobiles might be entirely contained and therefore safe in normal use, but they might pose a risk in severe accidents when the fuel tank ruptures. On the other hand, severe accidents and misuse will typically entail unusual risks even without nanotechnology, and it would be unreasonable to demand that every technology be entirely safe even when used unsafely.

Quite apart from possible risks, some critics argue that the benefits of nanotechnology themselves raise ethical issues, because they may not be available to everyone. Clearly, there are substantial political disagreements in society about whether socioeconomic inequality is unethical and, if so, under what circumstances and for what reasons. We can hope, as did the economists who participated in the societal implications conference mentioned earlier, that the free market will rapidly distribute the benefits more widely. However, this cannot be taken for granted, and in areas of public interest such as health care, governments may need to consider special efforts to maximize the distribution of benefits.

CONVERGING TECHNOLOGIES

Nanotechnology-related ethical issues are likely to become much more complex in the coming years, as progress in nanoscience makes possible more and more linkages to other technologies. This is has been described as *convergence at the nanoscale*:

> The phrase "convergent technologies" refers to the synergistic combination of four major "NBIC" (nano-bio-info-cogno) provinces of science and technology, each of which is currently progressing at a rapid rate: (a) nanoscience and nanotechnology; (b) biotechnology and biomedicine, including genetic engineering; (c) information technology, including advanced computing and communications; (d) cognitive science, including cognitive neuroscience. (Bainbridge 2002: ix; cf. Roco and Montemagno 2004)

Ultimately, this convergence will not consist merely of a collection of separate interdisciplinary fields that splice together different sciences and technologies in a piecemeal fashion. Rather, it will consist of a completely integrated set of theories and tools that provides comprehensive understanding and control of nature.

On one level, convergence is seen as the only way of continuing the rate of technological progress that humanity saw throughout the twentieth century. Demographers have noticed diminishing returns from medical science, as the rate of increase in the extension of the human life span markedly declined, but there is hope that nano-bio convergence can finally cure cancer and other diseases that have resisted a century's efforts and can begin to undo the degenerative effects of natural aging. It is believed that nano-info convergence can continue Moore's Law—the doubling of computer power at constant cost every 18 months—for another two decades, and this would mean computers having 8,000 times the power of today's machines.

But on another level, there is also realistic hope that entirely new things may be accomplished, even within the next very few years. Stimulated by defense needs, nano-bio-info convergence could result in sensors that can instantly identify even small quantities of any chemical substance or biological pathogen. While the patient is still in the doctor's office, this could mean definitive identification of an infection—viral or bacterial—allowing the doctor to begin proper treatment immediately.

For the first time, we have the real potential to create a host of cognitive technologies based on cognitive science supported by bio-nano-info innovations. These revolutionary changes raise profound ethical issues concerning privacy, equity, and possibly even the transformation of human nature.

PRACTICAL RESPONSES

Depending on the nature of the organization, any or all of the following four suggestions may help in dealing with ethical issues related to nanotechnology.

Strengthen professional ethics. No one company can go it alone, especially not small and medium-sized companies, but they can work together through professional organizations to set appropriate ethical standards, with periodic review to keep abreast of rapidly changing technical and social realities.

Delegate specific ethics-related duties. Ethics should be the explicit responsibility of high-level managers and of committees with wide representation, including labor as well as management and people representing a variety of functions in the company. This can be especially important in large corporations, where there is considerable division of labor and where communication problems can sometimes be severe.

Employ outside consultants. They would not make decisions, which are properly the task of corporate management, but would frame alternatives and inform the company about the wider debate currently in progress concerning particular issues of interest. Regularly, management and other representative employees could participate in seminars or similar activities facilitated by a consultant, sometimes primarily to communicate and reinforce standing company policies concerning ethical behavior and sometimes to inform participants about new issues that are on the horizon concerning nanotechnology ethics.

Encourage leaders to be well informed. Key personnel—including top management but also including other informal opinion leaders within the company and those who must maintain particular technical expertise—should individually inform themselves about the new possibilities for nanotechnology and for technological convergence at the nanoscale. Expect change, and look for signs of it. For a professional, incompetence is unethical.

The views expressed in this essay do not necessarily represent the views of the National Science Foundation or the United States.

REFERENCES

Bainbridge, William Sims. 1997. *Sociology.* New York: Barron's.

Bainbridge, William Sims. 2002. "Public Attitudes toward Nanotechnology," *Journal of Nanoparticle Research* 4: 561–570.

Bainbridge, William Sims. 2003. "Religious Opposition to Cloning," *Journal of Evolution and Technology* 13, retrieved September 1, 2004, from www.jetpress.org/volume 13/bainbridge.html.

Bainbridge, William Sims. 2004. "Social and Ethical Implications of Nanotechnology." In *Handbook of Nanotechnology,* ed. B. Bhushan. Berlin: Springer, 1135–1151.

Gauthier, David. 1986. *Morals by Agreement.* Oxford, England: Oxford University Press.

Homans, George C. 1974. *Social Behavior: Its Elementary Forms.* New York: Harcourt, Brace Jovanovich.

Kant, Immanuel. 1787. *A Critique of Pure Reason.* New York: Wiley [1900].

Luther, Wolfgang (ed.). 2004. *Industrial Applications of Nanomaterials—Chances and Risks.* Düsseldorf, Germany: Future Technologies Division, VDI Technologiezentrum.

Meridian Institute. 2004. *International Dialogue on Responsible Research and Development of Nanotechnology.* Alexandria, Virginia: Meridian Institute.

Moore, George Edward. 1951. *Principia Ethica.* Cambridge: Cambridge University Press.

Nietzsche, Friedrich Wilhelm. 1886–1887. *Beyond Good and Evil* and *The Genealogy of Morals.* New York: Barnes and Noble [1996].

Rawls, John. 1971. *A Theory of Justice.* Cambridge, Massachusetts: Harvard University Press.

Roco, Mihail C., and William Sims Bainbridge (eds.). 2001. *Societal Implications of Nanoscience and Nanotechnology.* Dordrecht, Netherlands: Kluwer.

Roco, Mihail C., and William Sims Bainbridge (eds.). 2003. *Converging Technologies for Improving Human Performance.* Dordrecht, Netherlands: Kluwer.

Roco, Mihail C., and William Sims Bainbridge (eds.). 2005. *Societal Implications of Nanoscience and Nanotechnology II: Maximizing Human Benefit.* Dordrecht, Netherlands: Kluwer.

Roco, Mihail C., and Carlo Montemagno (eds.). 2004. *The Coevolution of Human Potential and Converging Technologies.* New York: New York Academy of Sciences (Annals of the New York Academy of Sciences, volume 1013).

Royal Academy of Engineering. 2003. *Nanoscience and Nanotechnologies: Opportunities and Uncertainties.* London: The Royal Society.

Stark, Rodney, and William Sims Bainbridge. 1987. *A Theory of Religion.* New York: Toronto/Lang.

Stark, Rodney, and William Sims Bainbridge. 1996. *Religion, Deviance and Social Control.* New York: Routledge.

EPILOGUE

In late 1959 and again in 1983 Nobel Laureate and Caltech physics professor Richard Feynman delivered talks entitled, respectively, *There's Plenty of Room at the Botto* and *Infinitesimal Machines.*[1,2] Both offered questions and suggestions for research on what we now call "nanotechnology." By any name the subject matter was provocative and ahead of its time, so for many years these gems were not seen in the context of "real science" so much as novelties reflecting the rich, revered, idiosyncratic imagination for which Feynman was renowned—and still is, almost two decades after his death. Even in 1983 Feynman himself said "There is no use for these machines, so I still don't understand why I'm fascinated by the question of making small machines . . . ." Would that he had lived long enough to see the remarkable developments recounted in this book.

Although it is located here as an epilogue to the other chapters, the 1983 talk—a seminar at Jet Propulsion Laboratory—could as easily have stood at the other end as the book's foreword. A few comments on its presence and position are therefore in order.

Foremost, reprinting *Infinitesimal Machines* makes available the edited text of a fascinating speech whose distribution and appreciation are only a small fraction of that accorded Feynman's celebrated 1959 *Plenty of Room* speech. That earlier vision—and its exhortation to initiate research at microscale and below—is widely cited as the first public emergence of nanotechnology. The later talk, given and videotaped at JPL on February 23, 1983, was almost

---

1. Delivered before the American Physical Society, December 29, 1959, and transcript reprinted in Caltech's alumni magazine, *Engineering & Science*, February, 1960.
2. R. Feynman, *Infinitesimal Machinery*, IEEE Journal of Microelectromechanical Systems, 2, no. 1 (March 1993), 4–14. Copyright 1993, IEEE. The article is an edited transcript of the 1983 talk and is reproduced with the kind permission of Carl and Michelle Feynman, and the IEEE.

unknown beyond the fortunate attendees from the Caltech/JPL community until it was rediscovered, edited and published in 1993.[1,2]

Of course, any successor to *Plenty of Room* will unavoidably capture the attention of both the nano-community and Feynman fans. But, more specifically, the JPL talk sheds light on the ways of Feynman's thinking about things small, both mechanical and electronic, and the breadth of technology that his enthusiasms embraced. At the same time, because of skillful editing and "clean up" it preserves much of the Feynman personality and style that captivated students and researchers alike.

Feynman's interest in computing can be traced tothe World War II Manhattan Project at Los Alamos, where his application of innovative parallel computating methods tripled the throughput of arduous atomic bomb calculations perfomed by legions of calculation-punching staff who were the computers of the day. Fifteen years later at the time of *Plenty of Room*, computers were just transitioning to semiconductor technology, although vacuum tubes and even drum memory still remained much in use. By the time of his 1983 talk, the PC had arrived, Moore's law was being aggressively exercised, and Feynman's focus on computing had grown. Indeed *Infinitesimal Machinery* came in the middle of Feynman's year-long Caltech course "The Physics of Computation" which he taught with Carver Mead and John Hopfield. They had to hold it without him in 1981–1982, but with poor results, while Feynman battled cancer. His enthusiasm upon returning overcame their intention to abandon the course after the disastrous Feynmanless year. It was in this context that the JPL talk was given.[3]

---

1. Of the many people we have to thank for the lecture-to-article journey, a special debt is owed to *Journal* editor William Trimmer for discovering and pursuing the possibility and to Professor Steve Senturia, now retired from MIT, for yeoman editing that at once preserved both science and Feynman's charm.*NB:* the JPL lecture *is not* that sold under the title "Tiny Machines," a commercially distributed video of Feynman's 1984 talk at Esalen in Big Sur, California to an audience of artists and other non-scientists.
2. An illuminating analysis of the role of the two speeches and associated articles can be found in *Apostolic Succession* by Chris Toumey, "Engineering and Science" LXVIII, 1/2, 2005, available at

   http://pr.caltech.edu/periodicals/EandS/Esarchive-frame.html.
3. A deeper perspective of Feynman's involvement with computing can be found in *Feynman and Computation—Exploring the Limits of Computer*, ed. Anthony J.G. Hey (Philadelphia: Perseus Books, 1999). This is a collection articles by Mead, Hopfield and other notables who worked with Feynman. It also includes several of Feynman's own articles.

Placing *Infinitesimal Machines* as the final chapter reflects a belief that readers will appreciate it far more having first delved into contemporary nanotechnology. Moreover, the imagination and scope of Feynman's discussion are fitting symbols of the rich treasure of articles in this volume.

In that light and in closing, it bears mention that by motivating, organizing and editing these papers, Lynn Foster has made a major contribution to the field. This is no surprise, of course, to those who have watched him regularly encourage a thoughtful awareness of nanotechnology, in contrast to the hype that has persisted since the late 1990s. Initially working regionally to create the first Trade Study on Nanotechnology, he went on to contribute to and organize conferences with a national reach. Between the covers of this collection Lynn has brought together diverse authoritative perspectives that reveal the dimensions of nanotechnology, all the while written to be widely accessible. For the reporter who needs to get up to speed quickly, the scientist suddenly confronted with a research-driven business issue, or simply an inquisitive citizen wondering about all this "nano-hoopla," this volume offers a ready reference that is at once untainted by hyperbole and yet readable. For this we all owe Lynn Foster no small thanks.

—Michael Krieger, Caltech '63
Willenken Wilson Loh & Stris

Chapter 20

Infinitesimal Machinery

Richard Feynman[1]

Introduction of Richard Feynman by Al Hibbs—Welcome to the Feynman lecture on "Infinitesimal Machinery." I have the pleasure of introducing Richard, an old friend and past associate. He was educated at MIT and at Princeton, where he received a Ph.D. in 1942. In the War he was at Los Alamos, where he learned how to pick combination locks—an activity at which he is still quite skillful. He next went to Cornell, where he experimented with swinging hoops. Then, both before and during his time at Caltech, he became an expert in drumming, specializing in complex rhythms, particularly those of South America and recently those of the South Pacific. At Caltech, he learned to decode Mayan hieroglyphs and took up art, becoming quite an accomplished draftsman—specializing in nude women. And he also does jogging.

Richard received the Nobel prize, but I believe it was for physics and not for any of these other accomplishments. He thinks that happened in 1965, although he doesn't remember the exact year. I have never known him to suffer from false modesty, so I believe he really has forgotten which year he got the Nobel prize.

When Dick Davies asked me to talk, he didn't tell me the occasion was going to be so elaborate, with TV cameras and everything—he told me I'd be among friends. I didn't realize I had so many friends. I would feel much less uncomfortable if I had more to say. I don't have very much to say—but of course, I'll take a long time to say it.

---

1. This manuscript is based on a talk given by Richard Feynman on February 23, 1983, at the Jet Propulsion Laboratory, Pasadena, California. It is reprinted with the permission of his estate, Carl Feynman executor.

The author, deceased, was with the California Institute of Technology, Pasadena.

IEEE Log Number 9210135.

REVISITING "THERE'S PLENTY OF ROOM AT THE BOTTOM"

In 1960, about 23 years ago, I gave a talk called "There's Plenty of Room at the Bottom," in which I described the coming technology for making small things. I pointed out what everybody knew: that numbers, information, and computing didn't require any particular size. You could write numbers very small, down to atomic size. (Of course, you can't write something much smaller than the size of a single atom.) Therefore, we could store a lot of information in small spaces, and in a little while we'd be able to do so easily. And of course, that's what happened.

I've been asked a number of times to reconsider all the things that I talked about 23 years ago, and to see how the situation has changed. So my talk today could be called "There's Plenty of Room at the Bottom, Revisited."

As I mentioned in the 1960 talk, you could represent a digit by saying it is made of a few atoms. Actually, you'd only have to use one atom for each digit, but let's say you make a bit from a bunch of gold atoms, and another bit from a bunch of silver atoms. The gold atoms represent a one, and the silver atoms a zero. Suppose you make the bits into little cubes with a hundred atoms on a side. When you stack the cubes all together, you can write a lot of stuff in a small space. It turns out that all the books in all the world's libraries could have all their information—including pictures using dots down to the resolution of the human eye—stored in a cube 1/120 inch on a side. That cube would be just about the size you can make out with your eye—about the size of a speck of dirt.

If, however, you used only surfaces rather than the volume of the cubes to store information, and if you simply reduce normal scale by twenty-five thousand times, which was just about possible in those days, then the *Encyclopedia Britannica* could be written on the head of a pin, the Caltech library on one library card, and all the books in the world on thirty-five pages of the *Saturday Evening Post.* I suggested a reduction of twenty-five thousand times just to make the task harder, because due to the limitations of light wavelength, that reduction was about ten times smaller than you could read by means of light. You could, of course, read the information with electron microscopes and electron beams.

Because I had mentioned the possibility of using electron beams and making things still smaller, six or eight years ago someone sent me a picture of a book that he reduced by thirty thousand times. In the picture, there are letters measuring about a tenth of a micron across [*passes the picture around the audience*].

I also talked in the 1960 lecture about small machinery, and was able to suggest no particular use for the small machines. You will see there has been no progress in that respect. And I left as a challenge the goal of making a motor that would measure 1/64 of an inch on a side. At that time, the idea that I proposed was to make a set of hands—like those used in radioactive systems—that followed another set of hands. Only we make these "slave" hands smaller—a quarter of the original hands' size—and then let the slave hands make smaller hands and those make still smaller hands. You're right to laugh—I doubt that that's a sensible technique. At any rate, I wanted to get a motor that couldn't be made directly by hand, so I proposed 1/64 of an inch.

At the end of my talk, Don Glaser, who won the Nobel prize in physics—that's something that's supposed to be good, right?—said, "You should have asked for a motor 1/200 inch on a side, because 1/64 inch on a side is just about possible by hand." And I said, "Yeah, but if I offered a thousand-dollar prize for a motor 1/200 inch on a side, everybody would say 'Boy, that guy's a cheapskate! Nobody's ever going to do that.'" I didn't believe Glaser, but somebody actually did make the motor by hand!

As a matter of fact, the motor's very interesting, and just for fun, here it is. First look at it directly with your eye, to see how big it is. It's right in the middle of that little circle—it's only the size of a decimal point or a period at the end of a sentence. Mr. McLellan, who made this device for me, arranged it very beautifully, so that it has a magnifier you can attach—but don't look at it through the magnifier until you look at it directly. You'll find you can't see it without the magnifier. Then you can look through the magnifier and turn this knob, which is a little hand generator which makes the juice to turn the motor so you can watch the motor go around [*gives the McLellan motor to the audience to be passed around*].

WHAT WE CAN DO TODAY

Now I'd like to talk about what we can do today, as compared to what we were doing in those days. Back then, I was speaking about machinery as well as writing, computers, and information, and although this talk is billed as being about machinery, I'll also discuss computers and information at the end.

My first slide illustrates what can be done today in making small things commercially. This is of course one of the chips that we use in computers, and it represents an area of about three millimeters by four millimeters. Human beings can actually make something on that small a scale, with wires about six microns across (a micron is a millionth of a meter, or a thousandth of a millimeter). The tolerances, dimensions, and separations of some of the wires are controlled to about three microns. This computer chip was manufactured five years ago, and now things have improved so that we can get down to about one-half micron resolution.

These chips are made, as you know, by evaporating successive layers of materials through masks. [*Feynman uses "evaporating" as a generic term for all semiconductor process steps.*] You can create the pattern in a material in several ways. One is to shine light through a mask that has the design that you want, then focus the light very accurately onto a light-sensitive material and use the light to change the material, so that it gets easier to etch or gets less easy to etch. Then you etch the various materials away in stages. You can also deposit one material after another—there's oxide, and silicon, and silicon with materials diffused into it—all arranged in a pattern at that scale. This technology was incredible twenty-three years ago, but that's where we are today.

The real question is, how far can we go? I'll explain to you later why, when it comes to computers, it's always better to get smaller, and everybody's still trying to get smaller. But if light has a finite wavelength, then we're not going to be able to make masks with patterns measuring less than a wavelength. That fact limits us to about a half a micron, which is about possible nowadays, with light, in laboratories. The commercial scale is about twice that big.

So what could we do today, if we were to work as hard as we could in a laboratory—not commercially, but with the greatest effort in the lab? Michael Isaacson from the Laboratory of Submicroscopic Studies (appropriate for us) has made something under the direction of an artist friend of mine named Tom Van Sant. Van Sant is, I believe, the only truly modern artist I know. By truly modern, I mean a man who understands our culture and appreciates our technology and science as well as the character of nature, and incorporates them into the things that he makes.

I would like to show you, in the next slide, a picture by Van Sant. That's art, right? It represents an eye. That's the eyelid and the eyebrow, perhaps, and of course you can recognize the pupil. The interesting thing about this eye is that it's the smallest drawing a human being has ever made. It's a quarter of a micron across—250 millimicrons—and the central spot of the pupil is something like fifteen or twenty millimicrons, which corresponds to about one hundred atoms in diameter. That's the bottom. You're not going to be able to see things being drawn more than one hundred times smaller, because by that time you're at the size of atoms. This picture is as far down as we can make it.

Because I admire Tom Van Sant, I would like to show you some other artwork that he has created. He likes to draw eyes, and the next slide shows another eye by him. This is real art, right? Look at all the colors, the beauty, the light, and so forth—qualities that of course are much more appreciated as art. (Maybe some of you clever lit guys know what you're looking at, but just keep it to yourselves, eh?)

To get some idea of what you're looking at, we're going to look at that eye from a little bit further back, so you can see some more of the picture's background. The next slide shows it at a different scale. The eye is now smaller, and perhaps you see how the artist has drawn the furrows of the brow, or whatever it is around the eye. The artist now wants to show the eye to us on a still smaller scale, so we can see a little more of the background. So in this next slide, you see the city of Los Angeles covering most of the picture, and the eye is this little speck up in the corner!

Actually, all these pictures of the second eye are LANDSAT pictures of an eye that was made in the desert. You might wonder how someone can make an eye that big—it's two and one-half kilometers across. The way Van Sant made it was to set out twenty-four mirrors, each two feet square, in special locations in the desert. He knew that when the LANDSAT passes back and forth overhead, its eye looks at the land and records information for the picture's pixels. Van Sant used calculations so that the moment the LANDSAT looked at a particular mirror, the sun would be reflecting from the mirror right into the eye of the LANDSAT. The reflection overexposed the pixel, and what would have been a two-foot square mirror instead made a white spot corresponding to an area of several acres. So what you saw in the first picture was a sequence of overexposed pixels on the LANDSAT picture. Now that's the way to make art! As far as I know, this is the largest drawing ever made by man.

If you look again at the original picture, you can see one pixel that didn't come out. When they went back to the desert, they found that the mirror had been knocked off its pedestal, and that there were footprints from a jack rabbit over the surface. So Van Sant lost one pixel.

The point about the two different eyes is this: that Van Sant wanted to make an eye much bigger than a normal eye, and the eye in the desert was 100,000 times bigger than a normal eye. The first eye, the tiny one, was 100,000 times smaller than a normal eye. So you get an idea of what the scale is. We're talking about going down to that small level, which is like the difference in scale between the two-and-one-half-kilometer desert object and our own eye. Also amusing to think about, even though it has nothing to do with going small, but rather with going big—what happens if you go to the next eye, 100,000 times bigger? Then the eye's scale is very close to the rings of Saturn, with the pupil in the middle.

I wanted to use these pictures to tell us about scale and also to show us what, at the present time, is the ultimate limit of our actual ability to construct small things. And that summarizes how we stand today, as compared to how the situation looked when I finished my talk in 1960. We see that computers are well on their way to small scale, even though there are limitations. But I would like to discuss something else—small machines.

SMALL MACHINES—HOW TO MAKE THEM

By a machine, I mean things that have movable parts you can control, that have wheels and stuff inside. You can turn the movable parts; they are actual objects. As far as I can tell, this interest of mine in small machines is a misguided one, or more correctly, the suggestion in the lecture "Plenty of Room at the Bottom" that soon we would have small machines was certainly a misguided prediction. The only small machine we have is the one that I've passed around to you, the one that Mr. McLellan made by hand.

There is no use for these machines, so I still don't understand why I'm fascinated by the question of making small machines with movable and controllable parts. Therefore I just want to tell you some ideas and considerations about the machines. Any attempt to make out that this is anything but a game—well, let's leave it the way it is: I'm fascinated and I don't know why.

Every once in a while I try to find a use. I know there's already been a lot of laughter in the audience—just save it for the uses that I'm going to suggest for some of these devices, okay?

But the first question is, how can we make small machines? Let's say I'm talking about very small machines, with something like ten microns (that's a hundredth of a millimeter) for the size of a motor. That's forty times smaller than the motor I passed around—it's invisible, it's so small.

I would like to shock you by stating that I believe that with today's technology we can easily—I say *easily*—construct motors one-fortieth of this size on each dimension. That's sixty-four thousand times smaller than the size of McLellan's motor, and in fact, with our present technology, we can make thousands of these motors at a time, all separately controllable. Why do you want to make them? I told you there's going to be lots of laughter, but just for fun, I'll suggest how to do it—it's very easy.

It's just like the way we put those evaporated layers down, and made all kinds of structures. We keep making the structures a little thicker by adding a few more layers. We arrange the layers so that you can dissolve away a layer supporting some mechanical piece, and loosen the piece. The stuff that you evaporate would be such that it could be dissolved, or boiled away, or evaporated out. And it could be that you build this stuff up in a matrix, and build other things on it, and then other stuff over it. Let's call the material "soft wax," although it's not going to be wax. You put the wax down, and with a mask you put some silicon lumps that are not connected to anything, some more wax, some more wax, and then silicon dioxide or something. You melt out or evaporate the wax, and then you're left with loose pieces of silicon. The way I described it, that piece would fall somewhere, but you have other structures that hold it down. It does seem to me perfectly obvious that with today's technology, if you wanted to, you could make something one-fortieth the size of McLellan's motor.

When I gave the talk called "Plenty of Room at the Bottom," I offered a thousand-dollar prize for the motor—I was single at the time. In fact, there was some consternation at home, because I got married after that, and had forgotten all about the prize. When I was getting married, I explained my financial position to my future wife, and she thought that it was bad, but not so bad. About three or four days after we came back from the honeymoon, with a lot of clearing of my throat I explained to her that I had to pay a thousand dollars that I had forgotten about—that I had promised if somebody made a small motor. So she didn't trust me too much for a while.

Because I am now married, and have a daughter who likes horses, and a son in college, I cannot offer a thousand dollars to motivate you to make movable engines even forty times smaller. But Mr. McLellan himself said that the thousand dollars didn't make any difference—he got interested in the challenge.

Of course, if we had these movable parts, we could move them and turn them with electrostatic forces. The wires would run in from the edges. We've seen how to make controllable wires—we can make computers, a perfect example of accurate control. So there would be no reason why, at the present time, we couldn't make little rotors and other little things turn.

SMALL MACHINES—HOW TO USE THEM

What use would such things be? Now it gets embarrassing. I tried very hard to think of a use that sounded sensible—or semisensible—you'll have to judge. If you had a closed area and a half wheel that you turned underneath, you could open and shut a hole to let light through or shut it out. And so you have light valves. But because these tiny valves could be placed all over an area, you could make a gate that would let through patterns of light. You could quickly change these patterns by means of electrical voltages, so that you could make a series of pictures. Or, you could use the valves to control an intense source of light and project pictures that vary rapidly—television pictures. I don't think projecting television pictures has any use, though, except to sell more television pictures or something like that. I don't consider that a use—advertising toilet paper.

At first I couldn't think of much more than that, but there are a number of possibilities. For example, if you had little rollers on a surface, you could clean off dirt whenever it fell, and could keep the surface clean all the time.

Then you might think of using these devices—if they had needles sticking out—as a drill, for grinding a surface. That's a very bad idea, as far as I can tell, for several reasons. First, it turns out that materials are too hard when they are dimensioned at this small scale. You find that everything is very stiff, and the grinder has a heck of a job trying to grind anything. There's an awful lot of force, and the grinder would probably grind down its own face before it ground anything else. Also, this particular idea doesn't use the individualization that is possible with small machines—you can individually localize which one is turning which way. If I make all the small machines do grinding, I've done nothing I can't do with a big grinding wheel. What's nice about these machines—if they're worth anything—is that you can wire them to move different parts differently at different times.

One application, although I don't know how to use it, would be to test the circuits in a computer that is being manufactured. It would be nice if we could go in and make contacts at different places inside the circuit. The right way to do that is to design ahead of time places where you could make contacts and bring them out. But if you forgot to design ahead, it would be convenient to have a face with prongs that you could bring up. The small machines would move their little prongs out to touch and make contact in different places.

What about using these things for tools? After all, you could drill holes. But drilling holes has the same problem—the materials are hard, so you'll have to drill holes in soft material.

Well, maybe we can use these tools for constructing those silicon devices. We have a nifty way of doing it now, by evaporating layers, and you might say, "Don't bother me." You're probably right, but I'd like to suggest something that may or may not be a good idea.

Suppose we use the small machines as adjustable masks for controlling the evaporation process. If I could open and close these masks mechanically, and if I had a source of some sort of atoms behind, then I could evaporate those atoms through the holes. Then I could change the hole by changing the voltages—in order to change the mask and put a new one on for the next layer.

At the present time, it is a painstaking job to draw all the masks for all the different layers—very, very carefully—and then to line the masks up to be projected. When you're finished with one layer you take that layer off and put it in a bath with etch in it; then you put the next layer on, adjust it, go crazy, evaporate, and so on. And that way, we can make four to five layers. If we try to make four hundred layers, too many errors accumulate; it's very, very difficult, and it takes entirely too long.

Is it possible that we could make the surfaces quickly? The key is to put the mask next to the device, not to project it by light. Then we don't have the limitations of light. So you put this machine right up against the silicon, open and close holes, and let stuff come through. Right away you see the problem. The back end of this machine is going to accumulate goop that's evaporating against it, and everything is going to get stuck.

Well then, you haven't thought it through. You should have a thicker machine with tubes and pipes that brings in chemicals. Tubes with controllable valves—all very tiny. What I want is to build in three dimensions by squirting the various substances from different holes that are electrically controlled, and by rapidly working my way back and doing layer after layer, I make a three-dimensional pattern.

Notice that the silicon devices are all two-dimensional. We've gone very far in the development of computing devices, in building these two-dimensional things. They're essentially flat; they have at most three or four layers. Everyone who works with computing machinery has learned to appreciate Rent's law, which says how many wires you need to make how many connections to how many devices. The number of wires goes up as the 2.5 power of the number of devices. If you think a while, you'll find that's a little bit too big for a surface—you can put so many devices on a surface, but you can't get the wires out. In other words, after a while this two-dimensional circuit becomes all wires and no devices, practically.

If you've ever tried to trace lines in two dimensions to make a circuit, you can see that if you're only allowed one or two levels of crossover, the circuit's going to be a mess to design. But if you have three-dimensional space available, so that you can have connections up and down to the transistors, in depth as well as horizontally, then the entire design problem of the wires and everything else becomes very easy. In fact, there's more than enough space. There's no doubt in my mind that the ultimate development of computing machines will end up with the development of a technology—I don't mean my technology, with my crazy machines—but *some* technology for building up three-dimensional circuits, instead of just two-dimensional circuits. That is to say, thick layers, with many, many layers—hundreds and hundreds of them.

So we have to go to three dimensions somehow, maybe with tubes and valves controlled at small scale by machines. Of course, if this did turn out to be useful, then we'd have to make the machines, and they would have to be three-dimensional, too. So we'd have to use the machines to make more machines.

The particular machines I have described so far were just loose pieces that were moving in place—drills, valves, and so forth that only operate in place. Another interesting idea might be to move something over a surface or from one place to another. For example, you could build the same idea that we talked about before, but the things—the little bars or something—are in slots, and they can slide or move all over the surface. Maybe there's some kind of T-shaped slot they come to, and then they can go up and down. Instead of trying to leave the parts in one place, maybe we can move them around on rollers, or simply have them slide.

ELECTROSTATIC ACTUATION

Now how do you pull them along? That's not very hard—I'll give you a design for pulling. [*At the blackboard. Feynman draws a rectangular block with a set of alternating electrodes creating a path for the block.*] If you had, for example, any object like a dielectric that could only move in a slot, and you wanted to move the object, then if you had electrodes arranged along the slot, and if you made one of them plus, and another one minus, the field that's generated pulls the dielectric along. When this piece gets to a new location, you change the voltages so that you're always pulling, and these dielectrics go like those wonderful things that they have in the department store. You stick something in the tube, and it goes whshhhht! to where it has to go.

There is another way, perhaps, of building the silicon circuits using these sliding devices. I have decided this new way is no good, but I'll describe it anyway. You have a supply of parts, and a sliding device goes over, picks up a part, carries it to the right place, and puts it in—the sliding devices assemble everything. These devices are all moving, of course, under the electrical control of computer stuff below them, under their surfaces. But this method is not very good compared to the present evaporation technique, because there's one very serious problem. That is, after you put a piece in, you want to make electrical contacts with the other pieces, but it's very difficult to make good contacts. You can't just put them next to each other—there's no contact. You've got to electrodeposit something or use some such method, but once you start talking about electrochemically depositing something to seal the contact, you might as well make the whole thing the other way by evaporation.

Another question is whether you should use AC or DC to do the pulling: you could work it either way. You could also do the same thing to generate rotations of parts by arranging electrostatic systems for pulling things around a central point. The forces that will move these parts are not big enough to bend anything very much; things are very stiff at this dimensional scale.

If you talk about rotating something, the problem of viscosity becomes fairly important; you'll be somewhat disappointed to discover that if you left the air at normal air pressure in a small hole ten microns big, and then tried to turn something, you'd be able to do it in milliseconds, but not faster. That would be okay for a kit of applications, but it's only milliseconds. The time would be in microseconds, if it weren't for viscous losses.

I enjoy thinking about these things, and you can't stop, no matter how ridiculous things get, so you keep on going. At first, the devices weren't moving—they were in place. Now they can slide back and forth on the surface. Next come the tiny, free-swimming machines.

MOBILE MICROROBOTS

What about the free-swimming machine? The purpose is no doubt for entertainment. It's entertaining because you have control—it's like a new game. Nobody figured when they first designed computers that there would be video games. So I have the imagination to realize what the game here is: You get this little machine you can control from the outside, and it has a sword. The machine gets in the water with a paramecium, and you try to stab it.

How are we going to make this game? The first problem is energy supply. Another one is controlling the device. And if you wanted to find out how the paramecium looks to the device, you might want to get some information out.

The energy supply is, I think, fairly easy. At first it looks very difficult because the device is free-swimming, but there are many ways to put energy into the device through electrical induction. You could use either electrical or magnetic fields that vary slowly, generating EMFs inside.

Another way, of course, is to use chemicals from the environment. This method would use a kind of battery, but not as small as the device. The whole environment would be used—the liquid surrounding the device would be the source of a chemical reaction by which you could generate power. Or you could use electromagnetic radiation. With this method you would shine the light on the device to send the signal, or use lower frequencies that go through water—well, not much goes through water but light.

The same methods can be used for control. Once you have a way to get energy in—by electrical induction, for example—it's very easy to put digits or bits on the energy signal to control what the machine is going to do. And the same idea could be used to send signals out. I shouldn't be telling people at JPL how to communicate with things that are difficult to get at or are far away—this is far away because it's so small. You'll figure out a way to send the signals out and get them back again—and enhance the pictures at the end.

It's very curious that what looks obvious is impossible. That is, how are you going to propel yourself through the liquid? Well, you all know how to do that—you have a tail that swishes. But it turns out that if this is a tiny machine a few microns long, the size of a paramecium, then the liquid, in proportion, is enormously viscous. It's like living in a thick honey. And you can try swimming in thick honey, but you have to learn a new technique. It turns out that the only way you can swim in thick honey is to have a kind of an "S" shaped fin. Twisting the shape pushes it forward. It has to be like a piece of a screw, so that as you turn it, it unscrews out of the thick liquid, so to speak. Now, how do we drive the screw?

You always think that there aren't any wheels in biology, and you say, "Why not?" Then you realize that a wheel is a separate part that moves. It's hard to lubricate, it's hard to get new blood in there, and so forth. So we have our parts all connected together—no loose pieces. Bacteria, however, have flagella with corkscrew twists and have cilia that also go around in a type of corkscrew turn. As a matter of fact, the flagellum is the one place in biology where we really do have a movable, separable part. At the end of the flagellum on the back is a kind of a disc, a surface with proteins and enzymes. What happens is a complicated enzyme reaction in which ATP, the energy source, comes up and combines, producing a rotational distortion *[here, Feynman is using his hands to simulate a molecule changing shape and experiencing a net rotation];* when the ATP releases, the rotation stays, and then another ATP comes, and so forth. It just goes around like a ratchet. And it's connected through a tube to the spiral flagellum that's on the outside.

Twenty years ago when I gave my talk, my friend Al Hibbs, who introduced me today, suggested a use of small devices in medicine. Suppose we could make free-swimming little gadgets like this. You might say, "Oh, that's the size of cells—great. If you've got trouble with your liver, you just put new liver cells in." But twenty years ago, I was talking about somewhat bigger machines. And he said, "Well, swallow the surgeon." The machine is a surgeon—it has tools and controls in it. It goes over to the place where you've got plaque in your blood vessel and it hacks away the plaque.

So we have the idea of making small devices that would go into the biological system in order to control what to cut and to get into places that we can't ordinarily reach. Actually, this idea isn't so bad, and if we back off from the craziness of making such tiny things, and ask about a device that is more practical today, I think it is worth considering having autonomous machines—that is, machines that are sort of robots. I would tether the machines with thin wires—swallowing wires isn't much. It's a little bit discouraging to think of swallowing those long tubes with the optics fibers and everything else that would have to go down so the guy can watch the inside of your duodenum. But with just the little wires, you could make the device go everywhere, and you could still control it.

Even the wires are really unnecessary, because you could control the machine from the outside by changing magnetic fields or electric induction. And then we don't have to make the motors, engines, or devices so very tiny as I'm talking about, but a reasonable size. Now it's not as crazily small as I would like—a centimeter or one half of a centimeter—depending on what you want to do the first few times, the scale will get smaller as we go along, but it'll start that way. It doesn't seem impossible to me that you could watch the machine with X-rays or NMR and steer it until it gets where you want. Then you send a signal to start cutting. You watch it and control it from the outside, but you don't have to have all these pipes, and you aren't so limited as to where you can get this machine to go. It goes around corners and backs up.

I think that Hibbs's "swallowable surgeon" is not such a bad idea, but it isn't quite appropriate to the tiny machines, the "infinitesimal machines." It's something that should be appropriate for small machines on the way to the infinitesimal machines.

MAKING PRECISE THINGS FROM IMPRECISE TOOLS

These machines have a general problem, and that's the refinement of precision. If you built a machine of a certain size, and you said, "Well, next year I want to build one of a smaller size," then you would have a problem: you've only got a certain accuracy in dimensions. The next question is, "How do you make the smaller one when you've only got that much accuracy?" It gets worse. You might say, "I'll use this machine to make the smaller one," but if this machine has wobbly bearings and sloppy pins, how does it make an accurate, beautiful, smaller machine?

As soon as you ask that question, you realize it's a very interesting question. Human beings came onto the earth, and at the beginning of our history, we found sticks and stones—bent sticks and roundish funny stones, nothing very accurate. And here we are today, with beautifully accurate machines—you can cut and measure some very accurate distances.

How do you get started? How do you get something accurate from nothing? Well, all machinists know what you do. In the case of large machinery, you take the stones, or whatever, and rub them against each other in every which way, until one grinds against the other. If you did that with one pair of stones, they'd get to a position at which, no matter where you put them, they would fit. They would have perfectly matched concave and convex spherical surfaces.

But I don't want spherical surfaces—I want flat surfaces. So then you take three stones and grind them in pairs, so that everybody fits with everybody else. It's painstaking and it takes time, but after a while, sure enough, you've got nice flat surfaces. Someday, when you're on a camping trip, and everything gets boring, pick up some stones. Not too hard—something that can grind away a little bit, such as consolidated or weak sandstones. I used to do this all the time when I was a kid in Boston.

I'd go to work at MIT and on the way pick up two lumps of snow, hard snow that was pushed up by the snowplow and refrozen. I'd grind the snow all the way till I got to MIT, then I could see my beautiful spherical surfaces.

Or, for example, let's say you were making screws to make a lathe. If the screw has irregularities, you could use a nut that's breakable; you would take the nut apart and turn it backwards. If you ran the screw back and forth through the nut, both reversed and straight, soon you would have a perfect screw and a perfect nut, more accurate than the pieces you started with. So it's possible.

I don't think any of these things would work very well with the small machines. Turning things over and reversing and grinding them is so much work, and is so difficult with the hard materials, that I'm not really quite sure how to get increased precision at the very small level.

One way, which isn't very satisfactory, would be to use the electrostatic dielectric push-pull mechanism. If this device were fairly crude in shape, and contained some kind of a point or tooth that was used for a grinder or a marker, you could control the position of the tooth by changing the voltage rather smoothly. You could move it a small fraction of its own irregularity, although you wouldn't really know exactly what that fraction was. I don't know that we're getting much precision this way, but I do think it's possible to make things finer out of things that are cruder.

If you go down far enough in scale, the problem is gone. If I can make something one-half of a percent correct, and the size of the thing is only one hundred atoms wide, then I've got one hundred and not one hundred and one atoms in it, and every part becomes identical. With the finite number of atoms in a small object, at a certain stage, objects can only differ by one atom. That's a finite percentage, and so if you can get reasonably close to the right dimensions, the small objects will be exactly the same.

I thought about casting, which is a good process. You ought to be able to manufacture things at this scale by casting. We don't know of any limitation—except atomic limitations—to casting accurate figures by making molds for figures that match the originals. We know that already, because we can make replicas of all kinds of biological things by using silicone or acetate castings. The electron microscope pictures that you see are often not of the actual object, but of the casting that you've made. The casting can be done down to any reasonable dimension.

One always looks at biology as a kind of a guide, even though it never invents the wheel, and even though we don't make flapping wings for airplanes because we thought of a better way. That is, biology is a guide, but not a perfect guide. If you are having trouble making smooth-looking movable things out of rather hard materials, you might make sacs of liquid that have electric fields in them and can change their shapes. Of course, you would then be imitating cells we already know about. There are probably some materials that can change their shape under electric fields. Let's say that the viscosity depends on the electric field, and so by applying pressure, and then weakening the material in different places with electric fields, the material would move and bend in various ways. I think it's possible to get motion that way.

FRICTION AND STICKING

Now we ask, "What does happen differently with small things?" First of all, we can make them in very great numbers. The amount of material you need for the machines is very tiny, so that you can make billions of them for any normal weight of any material. No cost for materials—all the cost is in manufacturing and arranging the materials. But special problems occur when things get small—or what look like problems, and might turn out to be advantages if you knew how to design for them.

One problem is that things stick together by molecular attraction. Now friction becomes a difficulty. If you were to have two tungsten parts, perfectly clean, next to each other, they would bind and jam. The atoms simply pull together as if the two parts were one piece. The friction is enormous, and you will never be able to move the parts. Therefore you've got to have oxide layers or other layers in between the materials as a type of lubricant—you have to be very careful about that or everything will stick.

On the other hand, if you get still smaller, nothing is going to stick unless it's built out of one piece. Because of the Brownian motion, the parts are always shaking; if you put them together and a part were to get stuck, it would shake until it found a way to move around. So now you have an advantage.

At the end of it all, I keep getting frustrated in thinking about these small machines. I want somebody to think of a good use, so that the future will really have these machines in it. Of course, if the machines turn out to be any good, we'll also have to make the machines, and that will be very interesting to try to do.

COMPUTING WITH ATOMS

Now we're going to talk about small, small computing. I'm taking the point of view of 1983 rather than of 1960, and will talk about what is going to happen, or which way we should go.

Let's ask, what do we need to do to have a computer? We need numbers, and we need to manipulate the numbers and calculate an answer. So we have to be able to write the numbers.

How small can a number be? If you have *N* digits, you know the special way of writing them with base two numbers, that is, with ones and zeros. Now we're going to go way down to the bottom—atoms! Remember that we have to obey quantum-mechanical laws, if we are talking about atoms. And each of these atoms is going to be in one of two states—actually, atoms can be in a lot of states, but let's take a simple counting scheme that has either ones or zeros. Let's say that an atom can be in a state of spin up or of spin down, or say that an ammonia molecule is either in the lowest or the next lowest state, or suppose various other kinds of two-state systems. When an atom is in the excited state—a spin up—let's call it a "one"; a "zero" will correspond to spin down. Hereafter when I say a one, I mean an atom in an excited state. So to write a number takes no more atoms than there are digits, and that's really nothing!

REVERSIBLE GATES

Now what about operations—computing something with the numbers? It is known that if you can only do a few operations of the right kind, then by compounding the operations again and again in various combinations, you can do anything you want with numbers.

The usual way of discussing this fact is to have these numbers as voltages on a wire instead of states in an atom, so we'll start with the usual way. [*Feynman draws a two-input AND gate at the blackboard.*] We would have a device with two input wires *A* and *B*, and one output wire. If a wire has a voltage on it, I call it a "one"; if it has zero voltage, it's a "zero." For this particular device, if both wires are ones, then the output turns to one. If either wire is one, but not both, or if neither is one, the output stays at zero—that's called an AND gate. It's easy to make an electric transistor circuit that will do the AND gate function.

There are devices that do other things, such as a little device that does NOT—if the input wire is a one, the output is a zero; if the input wire is a zero, the output is one. Some people have fun trying to pick one combination with which they can do everything, for example, a NAND gate that is a combination of NOT and AND—it is zero when both input wires are ones, and one when either or both inputs are not ones. By arranging and wiring NAND gates together in the correct manner, you can do any operation. There are a lot of questions about branchings and so forth, but that's all been worked out. I want to discuss what happens if we try to do this process with atoms.

First, we can't use classical mechanics or classical ideas about wires and circuits. We have atoms, and we have to use quantum mechanics. Well, I love quantum mechanics. So, the question is, can you design a machine that computes and that works by quantum-mechanical laws of physics—directly on the atoms—instead of by classical laws.

We find that we can't make an AND gate, we can't make a NAND gate, and we can't make any of the gates that people used to say you could make everything out of. You see immediately why I can't make an AND gate. I've only got one wire out and two in, so I can't go backwards. If I know that the answer is zero, I can't tell what the two inputs were. It's an irreversible process. I have to emphasize this fact because atomic physics is reversible, as you all know, microscopically reversible. When I write the laws of how things behave at the atomic scale, I have to use reversible laws. Therefore, I have to have reversible gates.

Bennett from IBM, Fredkin, and later Toffoli investigated whether, with gates that are reversible, you can do everything. And it turns out, wonderfully true, that the irreversibility is not essential for computation. It just happens to be the way we designed the circuits.

It's possible to make a gate reversible in the following cheesy way, which works perfectly. [*Feynman now draws a block with two inputs, A and B, and three outputs.*] Let's suppose that two wires came in here, but we also keep the problem at the output. So we have three outputs: the *A* that we put in, the *B* that we put in, and the answer. Well, of course, if you know the *A* and the *B* along with the answer, it isn't hard to figure out where the answer came from.

The trouble is that the process still isn't quite reversible, because you have two pieces of information at the input, that is, two atoms, and three pieces of information at the output. It's like a new atom came from somewhere. So I'll have to have a third atom at the input [*he adds a third input line, labeled C*]. We can characterize what happens as follows:

Unless *A* and *B* are both one, do nothing. Just pass *A*, *B*, and *C* through to the output. If *A* and *B* are both one, they still pass through as *A* and *B*, but *C*, whatever it is, changes to NOT *C*. I call this a "controlled, controlled, NOT" gate.

Now this gate is completely reversible, because if *A* and *B* are not both ones, everything passes through either way, while if *A* and *B* are both ones on the input side, they are both ones on the output side too. So if you go through the gate forward with *A* and *B* as ones, you get NOT *C* from *C*, and when you go backward with NOT *C* as the output, you get *C* back again at the input. That is, you do a NOT twice, and the circuit, or atom, is back to itself, so it's reversible. And it turns out, as Toffoli has pointed out, that this circuit would enable me to do any logical operation.

So how do we represent a calculation? Let's say that we have invented a method whereby choosing any three atoms from a set of *N* would enable us to make an interaction converting them from a state of ones and zeros to a new state of ones and zeros. It turns out, from the mathematical standpoint, that we would have a sort of matrix, called *M*. Matrix *M* concerts one of the eight possible combination states of three atoms to another combination state of the three atoms to another combination state of the three atoms, and it's a matrix whose square is equal to one, a so-called unitary matrix. The thing you want to calculate can be written as a product of a whole string of matrices like *M*—millions of them, maybe, but each one involves only three atoms at a time.

I must emphasize that, in my previous example with AND gates and wires, the wires that carried the answer after the operation were new ones. But the situation is simpler here. After my matrix operates, it's the same register—the same atoms—that contain the answer. I have the input represented by *N* atoms, and then I'm going to change them, change them, change them, three atoms at a time, until I finally get the output.

THE ELECTRON AS CALCULATING ENGINE

It's not hard to write down the matrix in terms of interactions between the atoms. In other words, in principle, you can invent a kind of coupling among the atoms that you turn on to make the calculation. But the question is, how do you make the succession of three-atom transformations go bup-bup-bup-bup-bup in a row? It turns out to be rather easy—the idea is very simple. [*Feynman draws a row of small circles, and points often to various circles in the row through the following discussion.*]

You can have a whole lot of spots, such as atoms on which an electron can sit, in a long chain. If you put an electron on one spot, then in a classical world it would have a certain chance of jumping to another spot. In quantum mechanics, you would say it has a certain amplitude to get there. Of course, it's all complex numbers and fancy business, but what happens is that the Schrödinger function diffuses: the amplitude defined in different places wanders around. Maybe the electron comes down to the end, and maybe it comes back and just wanders around. In other words, there's some amplitude that the electron jumped to here and jumped to there. When you square the answer, it represents a probability that the electron has jumped all the way along.

As you all know, this row of sites is a wire. That's the way electrons go through a wire—they jump from site to site. Assume it's a long wire. I want to arrange the Hamiltonian of the world—the connections between sites—so that an electron will have zero amplitude to get from one site to the next because of a barrier, and it can only cross the barrier if it interacts with the atoms [of the registers] that are keeping track of the answer. [*In response to a question following the lecture, Feynman did write out a typical term in such a Hamiltonian using an atom-transforming matrix M positioned between electron creation and annihilation operators on adjacent sites.*]

That is, the idea is to make the coupling so that the electron has no amplitude to go from site to site, unless it disturbs the N atoms by multiplying by the matrix $M2$, in this case, or by $M1$ or $M3$ in these other cases. If the electron started at one end, and went right along and came out at the other end, we would know that it had made the succession of operations $M1$, $M2$, $M3$, $M4$, $M5$—the whole set, just what you wanted.

But wait a minute—electrons don't go like that! They have certain amplitude to go forward, then they come back, and then they go forward. If the electron goes forward, say, from here to there, and does the operation $M2$ along the way, then if the electron goes backwards, it has to do the operation $M2$ again.

Bad luck? No! $M2$ is designed to be a reversible operation. If you do it twice, you don't do anything; it undoes what it did before. It's like a zipper that somebody's trying to pull up, but the person doesn't zip very well, and zips it up and down. Nevertheless, wherever the zipper is at, it's zipped up correctly to that particular point. Even though the person unzips it partly and zips it up again, it's always right, so that when it's finished at the end, and the Talon fastener is at the top, the zipper has completed the correct operations.

So if we find the electron at the far end, the calculation is finished and correct. You just wait, and when you see it, quickly take it away and put it in your pocket so it doesn't back up. With an electric field, that's easy.

It turns out that this idea is quite sound. The idea is very interesting to analyze, to see what a computer's limitations are. Although this computer is not one we can build easily, it has got everything defined in it. Everything is written: the Hamiltonian, the details. You can study the limitations of this machine, with regard to speed, with regard to heat, with regard to how many elements you need to do a calculation, and so on. And the results are rather interesting.

HEAT IN A QUANTUM COMPUTER

With regard to heat: everybody knows that computers generate a lot of heat. When you make computers smaller, all the heat that's generated is packed into a small space, and you have all kinds of cooling problems. That is due to bad design. Bennett first demonstrated that you can do reversible computing—that is, if you use reversible gates, the amount of energy needed to operate the gates is essentially indefinitely small if you wait long enough, and allow the electrons to go slowly through the computer. If you weren't in such a hurry, and if you used idcal reversible gates—like Carnot's reversible cycle (I know everything has a little friction, but this is idealized)—then the amount of heat is zero! That is, essentially zero, in the limit—it only depends on the losses due to imperfections.

Furthermore, if you have ordinary reversible gates, and you try to drag the thing through as quickly as you can, then the amount of energy lost at each fundamental operation is one kT of energy per gate, or per decision, at most! If you went slower, and gave yourself more time, the loss would be proportionately lower.

And how much kT do we use per decision now? 10^{10} kT! So we can gain a factor of 10^{10} without a tremendous loss of speed, I think. The problem is, of course, that it depends on the size that you're going to make the computer.

If computers were made smaller, we could make them very much more efficient. It hadn't been realized previous to Bennett's work that there was, essentially, no heat requirement to operate a computer if you weren't in such a hurry. I have also analyzed this model, and get the same results as Bennett with a slight modification, or improvement.

If this device is made perfectly, then the computer could work ballistically. That is, you could have this chain of electron sites and start the electrons off with a momentum, and they simply coast through and come out the other end. The thing is done—whshshshsht! You're finished, just like shooting an electron through a perfect wire.

If you have a certain energy available to the electron, it has a certain speed—there's a relation between the energy and the speed. If I call this energy that the electron has kT, although it isn't necessarily a thermal energy, then there's a velocity that goes with v, which is the maximum speed at which the electron goes through the machine. And when you do it that way, there are no losses. This is the ideal case; the electron just coasts through. At the other end, you take the electron that had a lot of energy, you take that energy out, you store it, and get it ready for shooting in the next electron. No losses! There are no kT losses in an idealized computer—none at all.

In practice, of course, you would not have a perfect machine, just as a Carnot cycle doesn't work exactly. You have to have some friction. So let's put in some friction.

Suppose that I have irregularities in the coupling here and there—that the machine isn't perfect. We know what happens, because we study that in the theory of metals. Due to the irregularities in the positions or couplings, the electrons do what we call "scattering." They head to the right, if I started them to the right, but they bounce and come back. And they may hit another irregularity and bounce the other way. They don't go straight through. They rattle around due to scattering, and you might guess that they'll never get through. But if you put a little electric field pulling the electrons, then although they bounce, they try again, try again, and make their way through. And all you have is, effectively, a resistance. It's as if my wire had a resistance, instead of being a perfect conductor.

One way to characterize this situation is to say that there's a certain chance of scattering—a certain chance to be sent back at each irregularity. Maybe one chance in a hundred, say. That means if I did a computation at each site, I'd have to pass a hundred sites before I got one average scattering. So you're sending electrons through with a velocity v that corresponds to this energy kT. You can write the loss per scattering in terms of free energy if you want, but the entropy loss per scattering is really the irreversible loss, and note that it's the loss **per scattering, not per calculation step** [*heavily emphasized, by writing the words on the blackboard*]. The better you make the computer, the more steps you're going to get per scattering, and, in effect, the less loss per calculation step.

The entropy loss per scattering is one of those famous $\log_2$, numbers—let me guess it is Boltzmann's constant, k, or some such unit, for each scattering if you drive the electron as quickly as you can for the energy that you've got.

If you take your time, though, and drive the electron through with an average speed, which I call the drift speed, v_D (compared to the thermal speed at which it would ordinarily be jostling back and forth), then you get a decrease in the amount of entropy you need. If you go slow enough, when there's scattering, the electron has a certain energy and it goes forward-backward-forward-bounce-bounce and comes to some energy based on the temperature. The electron then has a certain velocity—thermal velocity—for going back and forth. It's not the velocity at which the electron is getting through the machine, because it's wasting its time going back and forth. But it turns out that the amount of entropy you lose every time you have 100% scattering is simply a fraction of k—the ratio of the velocity that you actually make the electron drift compared to how fast you could make it drift. [*Feynman writes on the board the formula:* $k(v_D/v_t)$.]

If you drag the electron, the moment you start dragging it you get losses from the resistance—you make a current. In energy terms, you lose only a kT of energy for each scattering, not for each calculation, and you can make the loss smaller proportionally as you're willing to wait longer than the ideal maximum speed. Therefore, with good design in future components, heat is not going to be a real problem. The key is that those computers ultimately have to be designed—or should be designed—with reversible gates.

We have a long way to go in that direction—a factor of 10^{10}. And so, I'm just suggesting to you that you start chipping away at the exponent.

Thank you very much.

Acronyms and Abbreviations

AFM	atomic force microscopy
AI	artificial intelligence
ANSI	American National Standards Institute
CMOS	complementary metal oxide semiconductor
CMP	chemical mechanical planarization or chemical mechanical polishing
CNTs	carbon nanotubes
CPSC	Consumer Products Safety Commission
CRADA	Cooperative Research and Development Agreement
CVD	chemical vapor deposition
DoD	Department of Defense
DOE	Department of Energy
DPN	dip-pen nanolithography
DRAM	dynamic random access memory
IBEA	Institute for Biological Energy Alternatives
IDD	implantable drug delivery
LEDs	light-emitting diodes
MEMS	microelectromechanical systems
MMS	molecular microswitch
MOSFET	metal oxide semiconductor field effect transistor
MRI	magnetic resonance imaging
MURI	Multi-University Research Initiative
NASA	National Aeronautics and Space Administration

NBIC	nano-bio-info-cogno
NBs	nanobelts
NCN	Network for Computational Nanotechnology
NDA	nondisclosure agreement
NDC	negative differential conductance
NEMS	nanoelectromechanical systems
NER	Nanoscale Exploratory Research
NIEHS	National Institute of Environmental Health Studies
NIH	National Institutes of Health
NIL	nanoimprint lithography
NIOSH	National Institute for Occupational Safety and Health
NIRTs	Nanoscale Interdisciplinary Research Teams
NIST	National Institute of Standards and Technology
NNCO	National Nanotechnology Coordination Office
NNI	National Nanotechnology Initiative
NNIN	National Nanotechnology Infrastructure Network
NNUN	National Nanofabrication Users Network
NSE	Nanoscale Science and Engineering
NSECs	Nanoscale Science and Engineering Centers
NSEE	Nanoscale Science and Engineering Education
NSET	Nanoscale Science, Engineering and Technology
NSF	National Science Foundation
NSP	Nanotechnology Standards Panel
NSRCs	Nanoscale Science Research Centers
NUE	Nanotechnology Undergraduate Education
NWs	nanowires
OSHA	Occupational Safety and Health Administration
OTT	office of technology transfer
PECVD	plasma-enhanced chemical vapor deposition
ROADMs	reconfigurable optical add/drop multiplexers
SAA	Space Act Agreement
SAM	self-assembling monolayer
SBIR	Small Business Innovation Research

SEM	scanning electron microscopy
SET	single-electron transfer
SOI	silicon on insulator
SPIONs	superparamagnetic iron oxide nanoparticles
SPM	scanning probe microscopy
STM	scanning tunneling microscope
STTR	Small Business Technology Transfer
SWNTs	single-walled nanotubes
TEM	transmission electron microscopy
URETI	University Research, Engineering and Technology
USPTO	U.S. Patent and Trademark Office

Index

D

informIT
informit.com/onlinebooks

THIS BOOK IS SAFARI ENABLED

INCLUDES FREE 45-DAY ACCESS TO THE ONLINE EDITION

The Safari® Enabled icon on the cover of your favorite technology book means the book is available through Safari Bookshelf. When you buy this book, you get free access to the online edition for 45 days.

Safari Bookshelf is an electronic reference library that lets you easily search thousands of technical books, find code samples, download chapters, and access technical information whenever and wherever you need it.

TO GAIN 45-DAY SAFARI ENABLED ACCESS TO THIS BOOK:

- Go to **http://www.awprofessional.com/safarienabled**
- Complete the brief registration form
- Enter the coupon code found in the front of this book on the "Copyright" page

If you have difficulty registering on Safari Bookshelf or accessing the online edition, please e-mail customer-service@safaribooksonline.com.